Raising Chickens

How to Sustain Your Garden's Ecosystem, Keep Away
Predators to Protect Your Flock, and Raise Healthy Chickens
for Homegrown Eggs

Claire Hennington

Table of Contents

Introduction

Have you ever watched a chicken pecking around in a yard and wondered what it would be like to have your own flock? Well, you are not alone. People all over the world are taking up the rewarding practice of raising chickens. But where do you start? Raising chickens can seem overwhelming if you have never done it before.

While Instagrammers make it look easy, there is a lot to consider, from choosing the right breed, building a secure coop, feeding and caring for your chickens to composting, and even dealing with potential predators. But there is no need to worry; this book was created to guide you through the whole process, step by step.

Just imagine stepping into your coop on a bright morning, and there, nestled in the nesting boxes, are a dozen—or more—eggs, fresh and warm and ready for breakfast! These eggs are not just any eggs; they are unprocessed, organic, and bursting with flavor. When you raise your own chickens, you can enjoy the delight of collecting eggs and eating chicken meat that you know is sourced from healthy and happy hens. Even better, you have a steady supply to share with family and friends.

Our modern era is marked by global warming—perpetuated by mass food processing industries—and serious environmental concerns. Sustainability is incredibly important, now more than ever.

Raising chickens aligns with the ethos of sustainable practices. You can reduce your carbon footprint by producing your own eggs and meat and give back to the earth by composting your flock's waste. Chickens also help with recycling kitchen scraps and contributing to healthier garden soil, creating a more self-sufficient and eco-friendly lifestyle.

Chickens are more than just egg producers; they are also charming and sometimes quirky pets. Their individual personalities and gentle clucks can bring a sense of calm and joy to your life. Whether you are spending a quiet afternoon in the yard or watching them scratch and peck their way through the garden, chickens have a unique way of capturing your heart.

A lot goes into starting your first flock, and raising chickens can seem overwhelming at first. Many beginner chicken keepers worry about how much time and initial investment goes into caring for these fascinating animals, which is precisely why I created this book with novices in mind.

This book is a comprehensive and practical guide for those looking to gain empowerment through sustainable living and chicken rearing. With my background in agricultural science, passion for sustainability, and a deep love for chickens, I have lovingly poured my knowledge and years of poultry farming experience into every chapter. Filling it with practical advice, and step-by-step coop building instructions to make your journey into chicken raising as smooth as possible.

Chicken raising has its rewards but it also comes with responsibilities. Chickens depend on us for food, shelter, and care. When you keep chickens your daily routine will involve chicken care, such as feeding, cleaning, egg collecting, and the occasional health check.

This book will help you assess whether you have enough space in your backyard or garden to accommodate a chicken coop and run; and we will even cover some legal talk including local zoning regulations, potential permits, how many feathered friends you are allowed to keep on your property, and so much more!

The initial phase of chicken-raising does involve some investment. You will need to purchase or build a coop, buy chicken feed, and provide bedding and other essentials. While the cost is not exorbitant, it is essential to budget for these expenses to ensure the well-being of your flock.

Poultry are vulnerable to various predators, including raccoons, foxes, and birds of prey. You will need to take measures to secure your coop and run to protect your chickens from harm. Consider whether you have the means and commitment to provide this level of security.

Chickens can be surprisingly vocal, especially when they are laying eggs. While their clucking is music to the ears of many chicken enthusiasts, you will need to consider how it might affect your neighbors. Open communication and the option of investing in noise-reduction strategies can help mitigate this concern.

After reflecting on the various considerations, it is time to make an informed decision about whether chicken raising is right for you. Remember that there is no one-size-fits-all answer. Your choice will depend on your lifestyle, values, and personal circumstances.

If you are ready to embrace the commitment, invest the time and resources, and provide a safe and nurturing home for these feathered friends, then chicken raising can be a fulfilling and rewarding adventure. This book is here to guide you every step of the way, from choosing the right breed to building a secure coop, ensuring their well-being, and enjoying the fruits of your labor. Chickens are more than just livestock; they can become an integral part of your life, offering companionship, entertainment, and connection.

One of the most delightful aspects of raising chickens is the companionship they provide. Chickens are curious and social creatures. They quickly learn to recognize their caregivers and will eagerly greet you with a chorus of clucks and coos when you approach. Unlike some livestock, chickens are relatively easy to tame and can become quite affectionate. You will find that spending time with your flock can be incredibly therapeutic and soothing.

Raising chickens can also provide a unique opportunity to connect with your food sources. In an age where much of our food comes from faceless, distant sources, knowing exactly where your eggs and meat come from can be empowering. You also gain a deeper appreciation for the effort and care that goes into food production.

If you have children or are an educator, raising chickens offers a hands-on and fun learning experience. Children can learn about the life cycle of chickens, from hatching to adulthood. They can witness their daily activities, from foraging for insects to dustbathing in the sun. This firsthand experience can foster a greater understanding of the natural world and the importance of responsible animal husbandry.

Chickens are more than just egg and meat producers; they are also skilled gardeners. Their scratching and pecking behavior helps control garden pests, reducing the need for chemical pesticides. Additionally, their manure is a valuable source of organic fertilizer, rich in nitrogen and other nutrients that can enhance soil health and plant growth. If you are an avid gardener, integrating chickens into your garden ecosystem can lead to healthier plants and increased yields.

There is a unique sense of accomplishment that comes with successfully raising chickens. As you watch your chicks grow into healthy adults and begin laying eggs or reaching a suitable size for meat production, you will feel a deep sense of pride in your role as a caregiver. Knowing that you have provided them with a safe and comfortable home, nutritious food, and proper care is incredibly satisfying.

Raising chickens is a practice deeply rooted in tradition and history. For centuries, people have kept chickens for their eggs, meat, and feathers. By continuing this practice, you are participating in a timeless agricultural tradition that connects you to countless generations of farmers and homesteaders who have cared for these birds. It is a way of preserving and honoring the skills and knowledge of our ancestors.

One of the most tangible rewards of raising chickens is the culinary delight they bring to your table. Fresh eggs from your own hens are a true delicacy. They have vibrant yolks, a rich flavor, and a texture that is noticeably different from store-bought eggs. Whether you enjoy them scrambled, poached, or as the star ingredient in baked goods, your homemade eggs will elevate your culinary creations.

If you are raising chickens for meat, you will have the satisfaction of knowing that the meat you are serving is of the highest quality. It is free from the antibiotics and hormones often found in commercial poultry and has been raised in humane conditions. This translates to meat that is not only healthier but also more flavorful and succulent.

As you embark on your chicken-raising journey, keep these joys and rewards in mind. They will serve as a source of motivation and inspiration as you encounter the inevitable challenges and learn the ropes of chicken keeping.

In the chapters that follow, we will delve deeper into the practical aspects of raising chickens. You will learn how to choose the right breed for your needs, create a secure and comfortable home for your flock, provide them with nutritious food and clean water, and keep them healthy and thriving. We will also explore more advanced topics, such as maximizing egg production, raising chickens for meat, and protecting your flock from predators.

By the time you finish reading this book, you will have gained the knowledge and confidence to raise your own chickens successfully. You will be well-prepared to enjoy the many benefits of chicken keeping, from fresh eggs and meat to the simple pleasure of watching these fascinating birds go about their daily lives. So, let's dive in and begin your chicken-raising adventure!

Chapter 1:

Is Chicken Raising Right for You?

Chickens are typically considered low-maintenance livestock and are naturally self-sufficient free-range eaters. When the initial startup investment in time and cost is over, the daily care of your chickens becomes fairly straightforward.

However, before embarking on the journey of chicken raising, you need to first evaluate if it is the right fit for you based on your commitment level, living situation, and personal preferences. From providing time and space for free-roaming, daily feeding, shelter provision, and flock protection, raising chickens requires careful time and initial investment.

Many beginner keepers have questions and concerns about the requirements of chicken raising, including if they have enough space, how to handle fertilized eggs, what happens when a hen stops laying, how to save money on feed, and questions about sustainability.

Other concerns include evaluating whether you meet the recommended space requirements, as well as making sure you are legally allowed to raise chickens based on your zoning restrictions.

Additionally, novice chicken farmers might question potential allergies or health issues that could be exacerbated by introducing chickens to your family.

Thankfully, these questions can be answered by careful research and proactive planning to ensure a lasting relationship between you and your flock. Whether seeking fresh eggs, sustainable living, or simply the joy of caring for these charming fowl.

Understanding the Commitment of Raising Chickens

Chickens are living creatures that depend on the attentiveness and dedication of their caretakers. Before jumping into the life of chicken farming, it is important to understand what it takes to ensure your hens thrive.

Assessing the Time and Effort Required

While chickens are relatively low-maintenance livestock, they still require daily care and attention, which includes tasks like feeding, cleaning, and general health checks. They also require a long-term commitment and if you decide to have a rooster or two you will likely be caring for some fuzzy and mischievous chicks.

Daily Feeding and Watering

In winter, spring, summer, and fall, chickens need fresh clean water daily, along with a variety of nutrient-rich foods. This might mean allowing your chickens to roam free, or providing them with daily fresh vegetables, select table scraps, and a consistent supply of high-quality chicken feed. If you live in an especially cold climate you may need to invest in a heat source to keep their drinking water from freezing. During the summer months, you may find that you need to refill their water more than once a day.

Cleaning the Coop and Maintaining Hygiene

In addition to daily watering and feeding, maintaining coop hygiene is one of the most vital ways to keep your birds healthy. Chickens can get various illnesses and diseases that are easily spread to the rest of your flock through droppings, shared feeders, and waterers. The best method for preventing illness is to thoroughly clean their coop, waterer, and feeder every day, while regular egg collection helps to prevent broody hens and reduce the occurrence of egg-eating predators (Chicken Whisperer Magazine, n.d.).

Regular Health Checks and Potential Veterinary Care

Keep an eye out for any signs of illness or injury in the coop. If you notice that something is 'off', identifying the issue early can be a lifesaver for your chicken. Unfortunately, finding a veterinarian familiar with poultry illnesses can be difficult and many chicken owners are left on their own. Even if you have access to a qualified veterinarian, you need to be aware of what symptoms to watch out for and what you should do if there is illness in your flock. Learning about chicken health and illnesses can be time-consuming and requires vigilant attention on your part. In certain situations, medicated feed or vaccinations may be necessary to ensure the health and safety of your flock.

Considering the Long-Term Commitment

Chickens have an average lifespan of five to ten years, which means you will need to seriously consider if you are ready for a long-term commitment. Some things you may want to consider are your lifestyle, work schedule, and the potential impact of the financial investment of raising chickens.

The Responsibility of Caring for Chickens Throughout Their Lifespan

Recognizing and preparing for different life stages, including raising chicks, introducing new birds, and managing older hens, is part of responsible poultry farming.

Chickens are hierarchical creatures and maintain their own system of flock politics, which means that pecking orders can be a real issue for many keepers. Typically, you can trust your flock to determine which hen ranks where in the chain of command. However, there are rare times when you may need to intervene. Hens on the bottom of the pecking order often get bossed around by those in higher ranks, and they often have to compete harder for food and may even get pecked, plucked, and sometimes attacked by other hens. How severe the bullying becomes varies, and depends on a range of factors, including the type of chicken breed.

If you have roosters with your hens, consider the option of fertilized eggs. The cycle of life is a beautiful thing and baby chicks are a great deal of fun. However, if you have a rooster running around, it might be more challenging to determine if an egg is fertile. Thankfully, this is easy enough to do with a bright light.

Another concern for many chicken keepers is protecting against predators. Depending on your location, some predators to be wary of are bears, bobcats, coyotes, foxes, raccoons, hawks, eagles, and even neighborhood cats and dogs. The threat of predators may determine whether or not you choose to keep your chicken free-range. Many keepers and farmers will train a guard dog to protect the flock, especially if they are free-range. However, regardless of how many precautions you take, some predators are sneaky and tenacious. Sadly, being prepared for unfortunate losses comes with the territory.

Predators are not the only reason for life losses in your flock. Just like any living creature you care for, chickens may pass away from old age, sudden illness, accidents, or even unknown causes. When a chicken passes away it may be emotionally difficult, especially for new keepers.

Regardless of the reasons you keep your flock—eggs, animal protein, or natural soil fertilization—coming to terms with the fact that chickens do not live forever is sad, so we need to be grateful for them every day for all that they do for us, from giving us companionship to providing healthy, sustainable sources of nourishment for us and our families.

Planning for Vacations or Times When You May Be Away From Home

There may be times when you are not able to take care of your chickens, such as during a vacation, an illness, or a change in your work schedule. So, you will need to make sure that your flock will have someone to care for them when you are unavailable.

It is not uncommon for first-time chicken owners to feel anxious about leaving their flock in the care of others. However, entrusting a family member, friend, or neighbor to take over chores and care will help alleviate any worries about the possibility of leaving your flock. One beneficial tip is to keep a daily care checklist for your designated "chicken-sitter" to follow along with an emergency contact method.

Evaluating Your Living Situation

There is more to owning chickens than daily care and dedication. Chickens are considered livestock, and some zoning laws prohibit the ownership of hens or roosters. Additionally, owning chickens may require quite a bit of space and can get smelly, especially in warmer climates. For those who also choose to keep roosters, noise levels may be an additional consideration.

There are many pros and cons to chicken rearing, and it may feel overwhelming to consider all the factors that come into play. However, there are many ways to make chicken raising work for your lifestyle, schedule, and budget. Caring for these unique and plucky creatures is well worth the consideration.

Assessing Space Requirements

For a healthy and well-adjusted flock, you need to have enough space in your backyard or garden to accommodate a chicken coop, and run if needed, and allow your chickens to roam free.

There are various breeds and the type of breed determines the temperament and size of the chicken (Lehr, 2022). Therefore, the amount of space you require to maintain happy hens will depend not only on the number of chickens but also on their size and breed temperament.

Minimum Space Requirements per Chicken

Square footage is important to keep in mind when considering coop size and free-roaming space. We will discuss individual chicken breeds and how to choose between them in later chapters. For now, the following information is meant to give you a general idea of how much space—at minimum—you can expect to dedicate to your coop.

A standard breed will typically need a coop with a minimum of four square feet per chicken and eight square feet of run space per chicken. However, if you choose to raise heavy breeds, then your coop will need at least eight square feet per chicken and a minimum of fifteen square feet of run space per heavy-breed chicken.

However, if your outdoor space is limited, fear not, your chicken-keeping dreams are not over! Bantam chickens are less than half the size of a standard chicken, which makes them a favorite choice among urban keepers as they require much less space compared to standard and heavy chicken breeds. These little fowls require a coop size of about two square feet per Bantam and a minimum run space of five square feet per Bantam.

Something else to keep in mind is the breed of the chicken, the more assertive the personality of your chicken the more space they will require. Giving assertive breeds more space will not only keep them happy but will also help prevent them from pecking and bullying lower-order hens (Lehr, 2022).

Checking Local Regulations

Research local regulations and restrictions on raising chickens to ensure you are compliant with the law. Zoning laws and permit requirements are something to keep in mind if you are planning to move or if your lifestyle requires you to move frequently.

Zoning Restrictions and Ordinances

Checking in with your local zoning and state regulations before adopting chickens will help you avoid fines and other possible legal repercussions. Typically, raising hens and roosters in areas that are already zoned for agriculture—such as a farm—is not an issue. However, keeping roosters might not be allowed in the neighborhoods surrounding farmland or more suburban and urban areas. Something else to keep in mind is whether you live in a rented property or a condo and have to comply with a Homeowners Association Committee. Additionally, your coop may need to pass local building codes and regulations.

Permit Requirements and Limitations

Depending on your location, you may be required to have a permit or have various limitations set on your chicken raising, such as:

- The number of chickens you can raise may be limited, depending on your plot size and zone.

- You might not be allowed to let your chickens roam free and might be required to keep your hens in a coop at all times.

- Some permits require annual renewal.

- Your chicken coop might have to comply with local building regulations.

- You might be required to build your coop a specific distance from the roads and homes.

- You might not be allowed to keep it except under specific circumstances.

- If you live in a residential area, you will likely not be allowed to slaughter your chickens for meat.

Considering Neighbors and Community

In addition to legal restrictions, it is good practice to consider the potential impact your chickens may have on your neighbors and community. Chicken can be noisy, dig up flower beds, antagonize local pets, get in the way of vehicles, or breed with other neighboring flocks potentially causing issues for other farmers or keepers.

Noise Levels and Potential Disturbance to Neighbors

As much fun as they are, chickens can be very noisy, especially if you have a rooster or two! The common phrase "up with the rooster's crow" is not just a metaphor. The truth is that roosters will crow whenever they feel like it at all hours of the day or night. If you are sensitive to noise or plan to keep your coop near your home, you might want to reconsider including them in your flock. It is also considerate to keep your neighbors in mind, not everyone takes kindly to a rooster exercising his lungs at all hours of the day or night—especially if you live in close proximity to your neighbors.

Something else to consider is not every neighborhood or village will be on board with the idea of having poultry running around all over the place, which will hinder your plans of allowing your flock to roam free.

Odor Control and Waste Management Strategies

Let's be honest, everybody poops, and your chickens will do it frequently. Once properly composted, chicken manure is one of the best fertilizers for your grass, vegetable gardens, and trees. However, reaping the benefits of this fantastic and free fertilizer means dealing with a stinky coop. The best way to keep a handle on odor control is a daily coop clean out and carrying out the fresh manure to your compost pile or bin.

Weighing the Pros and Cons of Chicken Raising

Raising chickens can be a rewarding process. Hens provide fresh eggs, sustainable living opportunities, and valuable life lessons for both children and adults. However, as wonderful and entertaining as chickens are, it is prudent to weigh the pros and cons before delving into poultry keeping. Understanding the full spectrum of the ups and downs that come with having a flock will help you make an informed and more confident decision about whether raising chickens aligns with your lifestyle and priorities.

Pros of Raising Chickens

Chickens are full of endless entertainment, many breeds are tame and like being held, cuddled, and stroked, making them quirky and friendly companions. Their clucking can be soothing, they are active and social, and provide an endless amount of feather-ruffling drama. In addition to their attitudes, these farm fowl can provide natural, sustainable, organic, and nutrient-rich food to our kitchen table.

Fresh Eggs and Meat Production

Good laying hens will provide you and your family with fresh eggs daily. Studies have shown that eggs from free-range chickens are packed with vitamins and minerals (Country Life Natural Foods, n.d.). Including nearly four to six times more vitamin D compared to store-bought eggs and chickens raised in confinement.

Roosters, hens that have stopped laying, and chicken breeds that are more suited for protein sources can provide a healthy and sustainable source of healthy meat. Raising chickens for consumption is a great alternative to buying mass-processed meat at the grocery store.

Sustainable and Self-Sufficient Living

Chickens are one of the best livestock choices for a sustainable and self-sufficient lifestyle. They provide a consistent source of compostable products such as manure and eggshells, and straw and hay from their coops. Compost with chicken manure is loaded with nitrogen, phosphorus, and potassium, which are vital to the healthy production of vegetable gardens and fruit trees (Chicken Whisperer Magazine, n.d.). Chickens are also great soil aerators, wherever they go, they scratch the soil in search of insects, plants, and tiny rocks. This loosening and disruption of the earth helps with healthy plant growth and soil vitality.

Chickens are also fantastic pest and weed control experts. Their natural love for grubs, beetles, and dandelions can help you manage pests and decrease the need to weed your vegetable garden during the growing seasons. Be warned: Do not let them in your vegetable garden or your feathered friends will dig up plants and eat any produce within reach!

For those looking to decrease their food waste, chickens may be one of the best additions to your kitchen as they will happily consume your scraps and table leftovers.

Educational Opportunities for Children

Chickens have so much character, each with a unique personality and quirkiness. Not only are they entertaining to watch, but they also make great learning opportunities for children.

Caring for chickens is educational and helps with instructing little ones to care for living creatures, in a way that teaches responsibility, compassion, and discipline. while it allows older kids to see and experience the various stages of the natural life cycle from birth, growth, reproduction, and death.

Cons of Raising Chickens

While there are many reasons why having chickens would be a great idea, there are also potential drawbacks and challenges of raising chickens to consider and be prepared for. Some of these drawbacks are easier to manage than others.

Time Commitment and Daily Care Responsibilities

Caring for chickens is not a one-and-done chore and you will need to clean their coop, water, and feed bins every single day. Chicken-proofing your garden bed and dodging chicken poop might be a necessity if your poultry are free-range.

Initial Setup Costs and Ongoing Expenses

Initial investments can vary depending on how many chickens you need, the available space, and your budget. However, chickens still require high-quality chicken feed every day. The cost of this can add up when considering that a healthy hen can consume two to four ounces of feed every day. Your flock may also need a heating source in the winter which, depending on the source and size of your coop, can add up costs to your electric bill.

Dealing With Potential Health Issues and Predators

Your flock will attract all sorts of natural predators. Depending on the types of predators you have in your region, it may be necessary to employ "proofing," against specific predators, to protect your coop. Proofing can be done in several ways and can typically be incorporated into the natural build of your coop. However, it is important to note that even the best proofing may not always be enough.

Chickens can become ill or injured and the fees for medication and treatment can add up, but chickens are not the only ones that can get sick. In addition to predators and health monitoring and prevention of your flock, your own health and immune system is something to consider before getting chickens. For some people, especially those with asthma or who are predisposed to allergies, keeping chickens can result in allergy flairs and potential respiratory health impacts (Country Life Natural Foods, n.d.).

Overcoming Common Objections and Concerns

Two of the most common worries beginner keepers have are time and cost. This is completely natural but there are plenty of ways to work around these concerns. The key is finding what works best for you and your family or lifestyle. Below are a few tips and tricks that may help you or even spur on some ideas.

Addressing Time Constraints

Effective time management strategies can help you fit chicken care into a busy lifestyle. A few common strategies include investing in automated equipment, creating a routine, and prioritizing tasks but ultimately, finding what works best for you and your flock might be a process of trial and error.

Time-saving tips for daily tasks:

- Prioritize tasks: Identify the most crucial coop tasks for the day and prioritize them accordingly. This means you can focus on what matters most while managing other commitments.

- Batch tasks: Combine similar tasks to save time. For instance, while feeding the chickens, use that time to check for eggs and observe their general well-being.

- Incorporate chickens into daily chores: Involve the chickens in your daily activities, such as letting them roam free while you garden or having them help with composting by giving them food scraps.

- Utilize lunch breaks or evenings: If your schedule allows, use your lunch break or evenings to take care of the chickens. This also provides an opportunity to unwind and connect with your feathered friends.

- Use time-saving tools and equipment: Choose time-saving tools, such as efficient cleaning tools for the coop or mobile coops that allow you to easily move the chickens to fresh grazing areas without much effort.

Strategies for planning and organizing chicken care:

- Create a schedule: Set a consistent daily routine for chicken care, incorporating feeding, cleaning, and other essential tasks. Having a schedule helps establish a sense of predictability and ensures that you allocate specific time for your chickens amidst other responsibilities.

- Have a checklist: A great way to keep track of chicken chores is to have a checklist visible on the fridge or by the door. This way it is easy for every household member to keep track of what has been done and what needs to be done.

- Delegate tasks: If possible, involve family members or housemates in caring for the chickens. Assign specific tasks to each person to share the workload and responsibilities.

- Plan ahead: Anticipate future busy periods in your schedule and make necessary adjustments to ensure that the chickens' needs are met. This could include having extra feed or arranging for someone to assist during your absence.

Managing Costs

Many beginners worry about the cost of keeping chickens. With careful planning and budgeting, it is possible to manage the expenses effectively. Suggestions include starting with a small flock, free-ranging your chickens, DIY-ing your coop, budget planning, and feeding efficiently.

DIY Coop Construction and Repurposing Materials

Building your coop rather than purchasing one can help you get the chicken coop of your dreams without breaking the bank. We will cover more on coop building in later chapters and provide you with plenty of resources, instructions, and advice.

If you are a bit more patient and savvy, then another great way to save on coop construction costs is to use repurposed materials. Great construction materials include:

- old barn lumber

- leftover tin roofing

- scrap lumber

- old screen doors

- wood materials from furniture

- old playhouses, sheds, or pool houses

- odds and ends from around the neighborhood

- pallets

A great place to find repurposed materials is online sites such as Facebook Marketplace, Craigslist, or even local chicken, farming, or second-hand groups.

Efficient Feeding and Waste Management Techniques

There are many ways to help mitigate yearly feed costs for your chickens. Here are three tried and true methods you can try out.

The first tip is to avoid giving your free-ranging chickens the option for "free-choice" feeding. Many chicken keepers and farmers choose free-choice feeding for their free-ranging flock, which means that the keepers maintain a constant supply of feed available for their chickens at all times. The assumption is that their birds automatically use the feed as supplement food when they need it. However, this might not be the best strategy, since your birds will be less likely to fill up on scavenged food like grubs, bugs, and greens. Instead, they will focus on the feed that is readily available to them.

To fix this issue, allow your birds to roam free first thing in the morning, then fill up feeders later on in the day for brooding hens. Your free rangers will return to their coops with full bellies and very little interest in their feeders. However, if you choose to follow this method make sure to remove the feeder at night, so that your chickens do not spoil their appetites with any midnight snacking.

Of course, if you live in less temperate climates and your chickens are unable to roam free during the winter, you will have to provide them with feed at all times.

The second tip is to grow your own chicken fodder—such as wheat grass—during the winter months. The best part is that growing fresh microgreens is a win-win for both you and your chickens! Growing wheatgrass fodder yourself is cost-effective, sustainable, and nutritionally conscious. Not only will you be able to feed your flock fresh greens every day, but you can also provide yourself and your family with healthy and organic microgreens all winter long.

The third tip is to ferment your feed during the cold winter months. Fermented feed is a great winter feeding option for two reasons. The first is that chickens' digestive systems absorb the nutrients more efficiently, which means they can eat less feed while gaining the same necessary nutrients and fat content. The second reason is that fermented feed is loaded with good probiotics that your free-range chickens might miss out on during long cold winters (Murana Chicken Farm, n.d.). Adding nutrient-dense food like this to your chickens' diet is especially important during months when they cannot forage.

Addressing Health and Safety Concerns

A few important questions and concerns many new chicken farmers have are what they should do if their chicken gets sick, whether their chickens can make them sick, and if it is necessary to wash fresh eggs.

Maintaining a healthy and safe environment for both chickens and humans mostly means following common sense practices, such as making sure your coop is clean, washing your hands regularly, looking for signs that your chicken might be sick, and following quarantine practices for chickens that are ill.

Disease Prevention and Biosecurity Measures

Keeping your coop clean is the best way to prevent illness and disease from spreading since most pathogens can be passed from bird to bird through fecal matter. Although it is not common among backyard and homestead chicken farmers, there are a few diseases that can be passed from your birds to their keepers. These include:

- Salmonella

- Listeria

- Campylobacter (jejuni)

- E. coli

Thankfully, these diseases can typically be avoided in humans by making sure you wash your hands after handling chicken droppings and collecting eggs, making sure your chickens always have clean fresh water, feeding your chickens high-quality feed, and cooking food properly.

If you notice any signs of illness in your chickens, you need to act immediately. The sooner you intervene, the more the risk of disease spreading to the rest of your chickens decreases. Below is a list of what you can do if you notice signs of illness in one of your birds (The Happy Chicken Coop, 2021b).

- Quarantine sick chickens in a crate or building away from your healthy flock.

- Wash your hands after handling a sick chicken.

- Do not share any feeders, waterers, or buckets between healthy birds and sick ones.

- Tend to your sick chicken after you have already taken care of your healthy birds.

- Make sure to find or call a veterinarian who can help treat your bird.

We also discuss how to identify, control, and treat illnesses in your flock in more detail in Chapter 6.

Safely Handling Chicken Eggs

The discussion on how to safely handle chicken eggs can bring up a pretty heated debate. There are two camps among chicken keepers, 'pro-washers' and 'pro-bloomers'. At the end of the day, the choice to wash your eggs or not is entirely up to you.

The argument against washing eggs is that running them under water removes the fragile protective 'bloom' or cuticle that effectively seals over the porous shell and protects the egg from bacteria. Additionally, keeping the bloom allows you to safely store your eggs at room temperature with no worries of spoilage. Eggs that still have their bloom can remain at room temperature for as long as a month before moving to the refrigerator. If immediately stored in the fridge, your eggs will stay fresh for up to six months!

Luckily, keeping your egg's bloom does not mean that you have to have dirty-looking eggs. If you would like to clean your eggs, gently wipe or brush away any dirt, droppings, or debris with a dry cloth or cleaning pad.

Those who do wash their eggs make the argument for personal aesthetics and fear of salmonella. However, it is important to note that when eggs are washed with water, they lose their bloom. This exposes them to bacteria—like salmonella—and the eggs must immediately be stored in the refrigerator. They will also have a shorter shelf life than eggs with their bloom intact. If you choose to wash your eggs, gently wash them with a scrubber, warm water, and soap. Be sure to use an all-natural, non-toxic, and food-safe soap. Do not let washed eggs soak with unwashed eggs as their protective barrier has been taken away.

In the end, whichever camp you fall into will depend on what works for you and makes you feel comfortable.

Interactive Exercise

Create a checklist to assess your readiness for raising chickens. Consider factors such as time availability, space requirements, potential challenges, and personal preferences. Reflect on each item and determine if you are prepared to take on the responsibilities of raising chickens.

Chapter 2:

Getting Started With Chickens and Choosing the

Right Chicken Breed

One of the most exciting parts of getting started with chicken raising is picking out your new chickens. There are so many breeds to choose from and there is no rule against mixing and matching different breeds to make up your unique flock!

Selecting the right breed is crucial since it helps you get the most out of your flock. The previous chapter mentioned that different breeds may have different space needs, and the importance of temperament, egg production, and availability when choosing the perfect breed for your flock. But there are a few other factors to consider when choosing a chicken breed including whether you want your chickens for eggs laying, meat, or both. Other factors include the climate you live in and whether you want to make chicken farming a business or simply use it to feed your family.

Differences in Chicken Breeds and Their Importance

The number of different chicken breeds is estimated to be in the hundreds, but no one can truthfully say for certain (The Happy Chicken Coop, 2022). Understandably, selecting the right breed may feel overwhelming. You would not want to start your farming journey off excited for loads of fresh eggs only to discover you have poor layers. Knowing your purpose for keeping chickens and understanding what makes each type of breed unique will help you narrow down your choices and make informed decisions. It is time to discover what breed has the ideal characteristics you are looking for in your flock.

Identifying the Best Chicken Breeds for Your Purpose

White, brown, black, spotted, freckled, and bespeckled, chickens come in all different colors and sizes. While there are innumerable breeds, each of them will fall into one of four categories. Dual purpose, egg-laying, meat, and heritage breeds (which we will cover in Chapter 9). Understanding each of these breeds will help you decide what breeds will be best suited to your needs and goals.

Dual-Purpose Breeds

If you want to raise chickens that are suitable for both egg and meat production, then dual-purpose breeds are a great choice. These hens tend to be productive egg layers—although not as prolific as egg-laying breeds—while also growing hefty enough to be used for their meat later in life. Select your dual-purpose breed carefully depending on whether you plan to use them for personal consumption or business.

As a sustainable homestead breed, these hens do not disappoint. However, take note that if you plan to sell your chicken eggs or meat, depending on the breed you select, dual-purpose hens might not produce eggs as prolifically as egg-laying breeds and do not grow as quickly or as large as meat breeds. When choosing dual-purpose hens for business purposes consider how large or busy you anticipate your business to be and if this breed will be able to keep up with the demand.

Egg-Laying Breeds

These hens are known for their high egg production, laying approximately four to six eggs per week. They will begin laying when they reach six months with a peak egg production age typically around two years old. These hens will continue to lay consistently until they reach retirement age and eventually stop laying.

While all egg-laying breeds are made for egg production, some are more prolific layers than others. Whether you plan to use these eggs for personal uses or sale, be sure to select the breed that will meet your production expectations.

Meat Breeds

These hens have been bred primarily for meat production and can pack on the pounds and grow rapidly. Some meat breeds can be ready for dinner within nine weeks (The Happy Chicken Coop, 2022).

Your reasons for raising meat chickens will not only determine the specific breed you choose but also the number of these hens you choose to raise.

Factors to Consider When Choosing a Breed

Aside from assessing your purpose and production expectations, a few other considerations when choosing a breed are regional climate conditions, your space requirements, different breed temperaments, and the noisiness of the breed.

Climate Adaptability

Consider the climate conditions of your region and choose a breed that can thrive in your specific environment. While most chickens are fairly adaptable to variable climates—as long as their keeper provides them with the appropriate shelter, food, and sanitary conditions—some breeds are naturally more suited to certain climates than others. Additionally, choosing a cold hardy breed will ensure that your chickens are suited to living in a colder climate and will save you from having to install a heating source in your coop in the winter.

Cold-hardy breeds for colder climates:

- Orpingtons

- Rhode Island Red

- Barred Rock

- Delawares

- New Hampshire Red

- Salmon Faverolles

- Araucana

- Australorp

- Marans (Caughey, n.d.)

Heat-tolerant breeds for hotter climates:

- Andalusian

- Leghorn

- Naked Neck

- White-Faced Black Spanish

- Ancona

- Minorca

- Penedesenca

- Sicilian Buttercup (Bri, n.d.)

Space Requirements

Consider the area where you plan to keep your flock, taking into consideration that you will need a coop and run space or free-ranging space, and select the most suitable breeds for the space you have.

Breeds More Suited to Free-Range Environments

Chickens are natural foragers with keen instincts for finding bugs, grass, and other snacks. No matter what breed you select, they will be happy and healthy to roam free. Some breeds are a little better at foraging and evading predators than others, which means that—depending on your environment and the natural predators in your area—selecting the right breeds can make all the difference to the longevity of your flock.

A few characteristics that make some chickens more successful free-rangers include attitude, size, coloration, and craftiness. Unfortunately, sweet docile chickens tend to get picked off by predators more easily. So, the best breeds for free-ranging and avoiding predators are the flighty and feisty types (Hamelman, 2021).

Independent and capable free-ranging breeds include:

- Ancona

- Barred Plymouth Rock

- Black chicken breeds

    - Ancona

    - Ayam Camani

    - Cochin

    - Jersey Giant

    - Sumatra

- Buckeye

- Egyptian Fayoumi

- Hamburg

- Jungle Fowl

- Lakenvelder

- Leghorn

- Old English Game Chicken

- Rhode Island Red (Hamelman, 2023)

Breeds That Thrive in Confined Spaces

We know that chickens are natural foragers and thrive outdoors; however, not every keeper has an environment or lifestyle conducive to free-ranging birds. Since docile breeds are not as adept at detecting danger and potential predators, they require a bit of extra protection. So, depending on your environment and lifestyle, some breeds may be safer kept in healthy confinement.

Breeds that are well suited to coop life, either part-time or full-time, include:

- Barbu D'Uccles

- Belgian D'Anvers

- Brahma

- Buff Orpington

- Cochin

- Delaware

- Easter Egger

- Golden Comet

- ISA Brown

- Plymouth Rock

- Polish

- Salmon Faverolles

- Silkie

- Sussex

- Wyandotte (Howell, 2022b)

Temperament and Handling

Temperament is something to consider especially if you have small children or if you are newer to chicken raising and do not feel comfortable with more flighty hens.

When you pick hens that suit your lifestyle, you contribute to the overall happiness and well-being of both your flock and yourself. If you are looking for a more hands-off approach to chicken raising like free-ranging, it is best to pick breeds that have temperaments suited to that lifestyle.

Temperament is especially important if you have a mix of different breeds in your flock. Make sure to keep docile chickens with docile chickens and feisty hens with other feisty hens, since chickens with bigger personalities can act out aggressively toward those who are more relaxed and friendly. Selecting breeds based on temperament can help you prevent bullying, acts of aggression, and injury.

Friendly and Docile Breeds for Families With Children

Chickens are famously unique and quirky. Some enjoy being pet, cuddled, and held by both keepers and their children. Other friendly hens provide multicolored eggs that thrill and fascinate children and grown-ups alike.

There are many sweet, gentle, and fluffy chicken breeds that families with children can happily enjoy. When planning to raise a flock, decide how much interaction you expect to have with your hens and what kind of breed temperament would be best suited to your family's needs, especially if you have small children.

A few great breeds to raise alongside children are:

- Australorp
- Bantams
- Brahma Chicken
- Buff Orpington
- Cochin
- Easter Egger
- Plymouth Rock
- Polish Chicken
- Silkie Bantam
- Speckled Sussex
- Wyandotte (Howell, 2022)

Independent and Less Human-Dependent Breeds

The breeds most suited for more hands-off caretakers are the free-rangers. These plucky hens know how and where to source their food during the day and return home to the coop at night.

While these breeds are independent it does not necessarily mean that they are unfriendly. Rhode Island Reds are a popular breed among farmers, homesteaders, and other keepers due to their independent, friendly, and laid-back personalities (Howell, 2022). Keep in mind that while your free-rangers might not be cuddlers, most of them are as friendly as they are independent.

Noise Levels

The only thing more important than a fresh egg haul to small space keepers is an acceptable noise level. Whether you are a backyard homesteader, suburban keeper, or urban gardener, the last thing you want is a complaint from a neighbor or to be woken up by clucking and crowing early in the morning.

The truth is that noise levels typically come down to the individual bird, even among generally 'quieter' chickens there may be a loud one in the bunch. Noise levels may depend on various factors such as the amount of space available and how much entertainment is provided for your chickens. Remember that a bored chicken will always be a noisy chicken.

Quiet Breeds Suitable for Urban or Suburban Environments

While all chickens make noise, some breeds are typically quieter and calmer than others. These breeds might be more suitable if you have specific noise restrictions and requirements. A few of these slightly quieter chicken breeds include

- Ameraucana

- Australorp

- Bantam

- Barred Plymouth Rock

- Brahma

- Buff Orpington

- Cochin

- Java

- Phoenixes

- Rhode Island Reds

- Salmon Faverolles

- Silkies

- Sussex Chicken

- Wyandottes (Dodrill, 2017)

More Vocal Breeds

Some breeds, like all variations of Orpingtons and Polish chickens, are known for making a lot of noise. If you live in a more suburban or urban area, you may want to steer clear of those two, along with roosters.

Roosters are loud regardless of their breed, so be sure to keep this in mind when purchasing chicks to add to your flock.

Popular Chicken Breeds for Egg Production and Meat

Picking out the correct chicken breed can be challenging and, in some cases, quite overwhelming. This is where the list below comes in. It offers a brief overview of a few of the most popular breeds to serve your purposes.

Egg-Laying and Dual-Purpose Breeds

A few popular breeds known for their egg-laying abilities are the Rhode Island Red, Leghorn, and the Sussex. Popular broiler breeds (chickens grown for meat consumption) include the Cornish Cross, Plymouth Rock, and the Jersey Giant.

Rhode Island Red

These hens can lay about 5-6 brown eggs per week (Oklahoma State University, 2021). Not only are they great layers, but they are friendly, independent, hardy, and assertive. However, they should not be kept with more docile breeds, since they tend to bully less assertive hens. Rhode Island Reds will also need some extra coop space and area to roam.

Leghorn

Recognized for their white eggs and high productivity, laying around 4-6 large white eggs per week (Oklahoma State University, 2021a). Unlike the Rhode Island Reds, this breed tends to be much more flighty and nervous. Leghorns also do not appreciate being held or picked up so if you are looking for a cozy cuddling companion, these ladies are not your best bet. However, if you are an experienced chicken keeper and do not mind skittish birds, these hens will make fantastic layers.

Sussex

These gentle giants are dual-purpose chickens and reliable egg layers, typically producing 4-5 large brown eggs per week (Oklahoma State University, 2021c). While they are quite large—with hens weighing in at 7-8 lbs. and roosters at 9 lbs.—these docile birds are easy to handle and incredibly curious. These chickens come in various color patterns which can add variety and fun to a full flock of Sussex.

Meat Breeds

Sometimes referred to as "broiler breeds" these chickens are bred to provide food for your family, whether they are dual purpose or strictly for meat production.

Cornish Cross

Typically seen in commercial farming, these chickens have been bred for one purpose only, meat production. Cornish Cross is known for its startling growth rate and high meat yield. They are not good egg layers and are typically ready to be plucked from the flock after 8-10 weeks of maturity. Unfortunately, these chickens are not recommended for breeding as their eggs are often infertile, and by the time they reach breeding age, they are considered too obese (Thrifty Homesteader Team, 2023). If you do choose this type of chicken, you will likely need to purchase new chickens every year. Additionally, these birds tend to develop multiple health complications due to their genetics. Do not keep these birds with more assertive or feisty hens as they will be subjected to harsh bullying.

Plymouth Rock

Considered one of the oldest breeds in America, this dual-purpose breed is known to produce great quality meat and possess good egg-laying ability; however, egg production typically slows a bit after they turn three. This breed is physically fit and healthy, some even live for up to 20 years (Oklahoma State University, 2021b). Plymouth Rocks have big personalities, are very curious and friendly, and some even appreciate a cuddle or two. Overall, while this breed needs space to roam and prefers to be free-range, they are typically easygoing.

Jersey Giant

This heavyweight breed is suitable for both meat and egg production. While they take a bit longer to reach full maturity, they do grow larger than most other breeds. Jersey Giants make good parents to their chicks, meaning less chick fostering and raising for you. This breed also tends to be very friendly, gentle, and even cuddly, making them great options for families with kids.

Sourcing and Acquiring Chicks or Adult Birds

So far, we have discussed what it takes to care for chickens, the various breeds that exist, and the importance of choosing the right birds for you. But where do you find the chickens and how do you know you are getting chicks from a reliable source?

Tips on Where and How to Buy Chickens

Whether you are sourcing your chickens from a breeder, hatchery, or feed store, it is wise to ensure that they adhere to and are certified by The National Poultry Improvement Plan (NPIP). This means that the chicks are hatched from a healthy and contagion-free flock.

When placing your order and selecting your flock, there are two purchasing options available for order: chicks or pullets.

Chicks are available for purchase just a few days after they have hatched and it will be your responsibility to raise them into maturity. On average, hens will begin laying eggs around 5-6 months old. There are a few advantages to buying chicks—aside from watching these adorable, innocent, and delightfully fuzzy chicks grow up. Namely, the chick will be more accustomed to socializing and being physically handled by humans as a mature bird. Chicks are also much cheaper to purchase, have a lower risk of illness due to environmental exposure in the hatchery, and have not yet learned any poor behavioral problems.

The disadvantages of purchasing chicks include the need for special care such as a safe enclosure and special food. Additionally, you have to wait for them to grow up before getting any eggs, and you also risk still ending up with a rooster you did not want—although, this is more of a concern when ordering from a breeder. Because of this, it is always good practice to order a few 'just in case chickens'.

Pullets, on the other hand, are hens that are already near or at laying age. Ordering pullets means that you do not have to worry about raising chicks and you will be in the egg business in no time. Additionally, you do not have to worry about getting an accidental rooster. However, there are a few downsides to purchasing pullets including, lack of bonding ability, higher price tag compared to chicks, limited breed variety, risk of being sold elderly hens, and high risk of contagions, parasites, and illness in your hens.

Another factor to consider when chicken shopping is how many chicks you need to purchase. The best way to decide is to consider what your individual needs are and then choose the number of chicks plus a few extra, just in case you get a rooster or an unfortunate event occurs during transportation. A good way to gauge how many chickens you will need is to decide how many eggs you go through per week. A healthy chicken will lay approximately 4-6 eggs per week (The Chicken Chick, 2016). This means that if you expect to have a dozen eggs every week you will need to order at least three hens, plus one or two extra.

Local Hatcheries and Breeders

You can order your chicks from hatcheries either online, by phone, or at a local hatchery. Ordering from a local hatchery is ideal because it is much safer for the chicks and reduces the risk of trauma, injury, or death during shipping. Hatcheries offer an extensive range of breeds that can be pre-ordered months in advance. So, it is easy to find out from the hatchery if they carry the type of chicken you are looking for and when they will begin taking orders.

A great perk of hatcheries is that they typically have very high biosecurity standards, some offer vaccinations for your chicks, and many provide the option for you to pick up your chicks locally for free. Additionally, most hatcheries offer chick sexing which ensures you are getting all the hens (or roosters) you order.

If you are looking for a specific chicken breed and want to find the highest quality bird possible, so you can compete in chicken shows, or just because you want to support people who specialize in your favorite breed, then you might consider sourcing your chicks from a breeder.

A great place to look for a breeder is on online breeders' associations and clubs. These associations will often have a list of legitimate, reputable, and local breeders for you to reach out to and order specific and harder-to-find breeds. You can also double-check to see if the breeder is NPIP certified. Keep in mind that not all breeders can provide sexing, so you could order a dozen chicks and unfortunately end up with five roosters. Additionally, since breeders are specialists, you are likely going to pay top dollar for your chicks, especially if they have blue-ribbon-winning chickens.

If you do not have any local hatcheries or breeders available to you, then you do have the option to source your chicks from a corporately owned feed store such as Tractor Supply Company (TSC) or an independently owned feed store (The Chicken Chick, 2016).

Researching Local Hatcheries and Breeders

Be careful when searching and making orders online. Not all companies that sell chicks online are hatcheries or breeders. Instead, they are more like chicken dealers, who take orders from individuals and then send those orders to hatcheries and breeders all over the country, who then ship the chicks to you. This can be a dangerous journey for your fragile chicks as it may take days without food, water, or temperature regulation before they reach your doorstep. Therefore, when shopping online, make certain that you know where your chicks will be shipped from and whether or not the company you are ordering from is a legitimate hatchery or breeder, not just a middleman.

Visiting Hatcheries or Breeders in Person

It is always a better option to collect your chicks in person rather than having them shipped; however, this might not always be feasible. Therefore, make sure to communicate with your hatchery or breeder so that you will be home when your chicks are delivered.

Online Sources

You can also explore online platforms, websites and social media; Facebook groups are especially great options. While the sale of animals goes against Facebook's policies, you can network and meet people who are either breeders themselves or can direct you to a reputable breeder or seller in your area. Facebook groups can be local, breed-specific, or simply about general care, and are great resources to have in your pocket for information, comradery, or even indirectly sourcing your chickens.

Other online platforms include classified ads websites such as Craigslist. In general, these can be helpful resources in finding supplies such as coops, feeders, waterers, and other materials for a DIY coop. It is also possible to discover breeders and sellers through online classifieds; however, use caution when purchasing birds from such sources, since these might not always be reputable, or could even be a scam.

Research Online Sellers and Marketplaces

When deciding where to purchase your chickens, research, and knowledge is key. There is nothing more disheartening than starting your chicken-keeping dream with sick and failing chickens. Doing a quick Google search to confirm the reputation and legitimacy of a seller cannot hurt.

Read Reviews and Customer Feedback

Before purchasing your chickens online look for any previous customer reviews, and read recommendations from online chicken forms and groups. Make a list of reputable hatcheries, breeders, and farming supply stores to determine which is best for you.

Local Agricultural Fairs and Exhibitions

A great way to connect with breeders and see different chicken breeds in person is by attending local fairs and exhibitions. These events can also be highly educational and beneficial networking experiences. Some farmers and breeders may even have chicks available for purchase. Unfortunately, even though these birds may appear healthy, they might be carriers for some pretty nasty diseases that can be passed along to the rest of your flock. So be cautious when purchasing from these fairs.

Some of these events have a swap meet section, use extreme caution when purchasing or swapping chickens at such an activity. Due to the intermingling aspect of a swap meet, it is very easy for animals carrying contagions to pass along illness to all the other animals at the event and eventually, your flock. If you do choose to purchase a bird from a swap meet, practice quarantine procedures with the new animal the same way you would if one of your chickens were sick. Ensure your new bird is not showing any signs of illness and have a fecal sample sent to a lab or your veterinarian to test for any possible pathogens before you introduce it to your flock.

Locating Agricultural Fairs and Exhibitions in Your Area

You can locate agricultural fairs and exhibitions by asking local farmers, online groups, or other fellow chicken enthusiasts. Another way to discover these fairs is to check your local newspaper for upcoming events.

Networking With Breeders and Fellow Chicken Enthusiasts

Networking is one of the best things you can do. Having a group of individuals with the same interests and goals as you can be incredibly fulfilling. Networking with other chicken enthusiasts can help you make friends, build resources, learn, and be the best chicken keeper you can be!

Selecting Healthy Chicks or Adult Birds

It is easy to look at an arm full of yellow fluffy peepers and want to take them all home. Unfortunately, not every little ball of fluff is a healthy one and adults tend to be more susceptible to illnesses due to length of exposure. Make sure you purchase healthy birds to ensure the health and safety of yourself and your entire flock. If you notice any signs of disease or illness in a chick, pullet, or grown chicken do not purchase it.

Look for these signs:

- any discharge from nasal passage or eyes

- blocked nasal vent

- wheezing

- lethargy

- being hunched in a ball

- self-isolation

- inability to move after gentle coaxing

- dull looking eyes

- rough, dull, or unkept-looking feathers (The Happy Chicken Coop, 2022)

Introducing New Chickens to an Existing Flock

What happens if you have an established flock and want to add in a few extra chickens? Keep in mind that chickens have a pretty harsh pecking order and new birds can carry diseases. To avoid any issues, there are a few best practices to follow when introducing new friends to your flock.

Always quarantine new birds before bringing them into your flock to make sure that they are disease-free. When it comes to the health of your flock, it is better to be safe than sorry.

Make sure that the breeds you choose are compatible. Remember that not all chicken breeds get along or play nicely together. Avoid physical harm to your new chicken by making sure you are not sending a gentle lamb into a pack of feathered wolves.

Try to introduce multiple new chickens at once, rather than a single chicken. Imagine being the only new kid in a school full of clicks and hierarchies. Chances are, you will feel alone and even get bullied. But if you came into the school with a large group of new kids and you all knew each other, your transition to your new school would go much smoother. The same concept goes for new chickens: There is safety in numbers.

Do not introduce birds that are still too young. If your chickens are younger than 8-12 weeks, then they may not be big enough to defend themselves from bullying. Make sure that you wait until your chicks are big enough to compete with the other adult hens in the coop. Chicks that you have raised yourself can also transition to coop life much easier as adults since they are already used to you and your care.

Make sure that you have enough coop space to accommodate the new additions to the flock. Remember each breed requires a certain minimum amount of space. So make sure to keep your flock's space requirements in mind.

Switch up the territory. In other words, try to reduce pecking order upheaval by bringing the flock to your new bird and allowing them to get used to each other on neutral turf—so to speak. Another way you can switch things up is by moving your whole flock to a temporary holding area, while the new chicken explores the coop without bullying or worry. Then slowly begin adding flock members back to the coop.

Provide your chickens with some entertainment. This may distract them from the newcomer and allow a more peaceful transition period. Extra distractions could be anything from adding a hay bale stuffed with treats to scratch and peck out, hanging a cabbage, or giving your birds some veggies to forage through. Even adding toys like a treat ball or a swing, could serve as enough of a distraction.

Add an extra watering and feeding station. Older and more established hens tend to crowd out the lower ranks, so your newbie will be at the bottom of the pecking order, which means they will likely not have much opportunity to eat or drink until the acclimation period is over. Adding extra feed and water stations will allow your new chicken to have more space and opportunity to get the food and water they need.

Lastly, make sure to keep a careful eye on your new hen and check in on them as much as possible. This way, if you see that the new addition is being ganged up on, you will be able to take action and hopefully prevent any serious injury.

Interactive Exercise

Create a breed selection checklist that includes important factors such as purpose, climate adaptability, space requirements, temperament, and noise levels. Use this checklist to evaluate different chicken breeds based on your specific requirements and narrow down your options.

Chapter 3:

Building Your First Coop

Building your first chicken coop can be an exciting, rewarding, and occasionally daunting experience. Making sure your coop is up to spec provides a comfortable home for your feathered friend and opens the door to the joys of poultry keeping. However, a well-designed and constructed chicken coop is about more than providing a cozy space, it is also about ensuring the safety, health, and happiness of your flock. Learning about predator-proofing, the basics of coop design, suitable materials, important elements, and strategies for building your coop are all part of the building process.

Basics of Chicken Coop Design and Construction

What if you want to skip the building hassle and simply purchase a coop? Regardless of whether you choose to buy or build, you will need to consider the important elements that make a good coop including; size, ventilation, lighting, roosts, nest boxes, shade, run space predator protection, mobility, and even dust bath spaces.

Before you can dive into building plans, construction, or online shopping, it is important to be fully aware of all the different kinds of coops and how to spot a good one.

Additionally, when looking around for the best coop or when building, ask yourself:

- How large does the coop need to be for my chickens?

- Will I be free-ranging my birds and if not, do I have enough space for a run?

- Where in my yard will I be placing my coop?

- Are there enough nesting boxes?

- How many roosts should I have and what size do they need to be?

- How can I reinforce my coop for predators? (Arcuri, 2019)

Understanding Coop Design Principles

However, before you can begin looking into these aspects, you must first determine whether you require a mobile coop or a stationary one.

The type of coop you choose for your flock will depend entirely on whether they will be living inside full-time with access to a run or free-range space, or have access to large pastures and fresh ground to forage regularly.

If you plan on having your chickens in the pasture, you will need a movable coop—sometimes referred to as a 'chicken tractor'—which is capable of transporting your chickens from one portion of pasture to another. However, if you plan on keeping your chickens in one spot then a stationary coop is all you need.

A standard stationary coop typically includes an indoor enclosure. If you do not plan on allowing your chickens to roam free then they will need an attached, fenced, and covered run space. Adequate run space is vital to their overall health and happiness.

The design of a chicken tractor will depend largely on its usage. If you plan on keeping a few chickens as a permanent home, then you will need both an indoor coop space as well as adequate enclosed field forage space.

In both cases of coop or tractor, your chickens will still need roosting and nesting boxes, and space to forage, roam, and release energy. The size of your flock will determine the overall space requirements as well as the quantity of nesting boxes and roosting spots.

Coop Size and Space Requirements

In Chapter 1, we briefly touched on the subject of determining coop size based on the number of birds and breed types. However, much more goes into determining the size of your coop, and generally, a coop should be a little bit bigger in size than you initially planned, as it will offer you more flexibility and options to expand your flock in the future (Arcuri, 2019).

Here are a few more pointers to help you calculate how big your coop should be. Please note that these recommendations are given assuming standard-size chickens with relatively moderate personalities. If you have larger birds or big personality birds then you will need to accommodate your breed by adding extra square footage per chicken.

- If you plan on having a movable coop with plenty of pasture for your chickens to forage in, you will need approximately 10 square feet per chicken (Arcuri, 2022).

- If you plan on having free-range birds, you will need approximately 4 square feet per chicken available inside the coop with about 8 square feet of run space per bird (Lehr, 2022).

- However, if you have to keep your chickens indoors for the winter—or any other portion of the year—each chicken will require about 10 square feet inside the coop (Lehr, 2022).

So how much space do you really need?

- If you have six chickens, then you will need approximately 24 square feet of coop space and 48 square feet of run space at a bare minimum for standard breed chickens.

- If you have ten chickens then you will need 40 square feet of coop space and a minimum of 80 square feet of run space.

- If you have a 'chicken tractor' and six chickens then you will need approximately 60 square feet.

You can calculate square footage easily by determining how many chickens you plan to have and how much space they require. For example, if you have ten chickens, they will need 4 square feet of space each. So you would simply multiply the number of chickens (10) by their space requirement (4 feet). 10 x 4 = 40, so your coop would need to have 40 square feet of space (Lehr, 2022).

Ventilation and Airflow For Optimal Air Quality

A good coop will provide proper ventilation, which helps prevent your flock from contracting nasty respiratory diseases and helps you deal with odor control. However, it is important to strike a balance since the coop should not be too drafty either.

Ideally, your ventilation setup should be adjustable to maintain proper airflow and bird health through all four seasons. If you live in hot and dry climates, consider building a coop with higher ceilings, which will add extra air flow and keep the coop cooler (Smith, 2020b).

Lighting Considerations to Encourage Egg Production

Chickens require around 12 to 14 hours of daylight to be at peak egg production (Cocoon Chicken Coops, 2022). During the summer months, this is usually not a problem. However, during the winter months, when the days are shorter, your chickens may only receive 8 to 10 hours of daylight, which slows down egg production. So instead of your hens laying four eggs a week, you might only get two or maybe three per bird. Because of this, some farmers and homestead egg enthusiasts, choose to set up artificial lighting for their chickens in the colder months. If you choose this route, make sure that you keep the lights on a timer so that your chickens are not getting more than 12 hours of light. After all, sleep is just as important for their health as it is for yours. Well-rested hens are also better egg layers.

If you do add artificial lighting, make sure that the light source is far enough away from the chickens so that they cannot get to it or break it. Ensuring that you have a chicken-safe light is key to preventing injury, death, or even fires.

If you do not mind a slowdown in egg production during the winter months. then simply make sure that your chickens have plenty of natural light. You can accomplish this by adding some sturdy insulated windows to your coop for your chickens to get all the light they need.

Choosing a Suitable Location

Choosing the right location might be a challenge, especially if you have limited backyard space. You also need to keep in mind access to sunlight, protection from the elements, access to roaming space, and distance from your home.

Evaluating Available Space in Your Backyard or Garden

Take a good look around your yard and pick a spot that will provide both natural light during the day and shade during the hottest time of the day. A good place to put a coop is under a tree or in another sheltered location such as next to a barn or shed that will provide afternoon shade. If you do not have such a space available, just make sure to build in some extra shade and ventilation to your coop design. Wherever you decide to build, make sure that you have enough space to do so.

A good way to decide if you have enough space in your yard is to grab a tape measure, some rope or string, and a few stakes or sturdy sticks to measure out your coop and run space in the spot you have chosen. This will help you visualize exactly what the coop will look like and how much space it will take up.

Considering Proximity to Your House

Proximity is key. You would not want your chicken coop to be so close to the house that you avoid opening the windows in the summer, but not so far away that you do not have easy access. A distance of at least 10 to 200 feet away from your home should suffice (Smith, 2020a). Keep in mind that you will have to tend to your chickens in various weather conditions, so the overall distance will depend on your convenience and what works best for the outdoor space you have.

Assessing the Ground Levelness for Stable Construction

Another thing to keep in mind when selecting a spot is to make sure that you have a relatively flat location to build your coop. This is important for the long-term stability and safety of your structure. If you do not have a flat location, consult a professional builder to assess the options.

Choosing the Right Materials for Your Coop

When selecting your materials ensure that you are always keeping coop security, sustainability, and maintenance a priority. While material selection may vary depending on whether you plan to upcycle, DIY, or purchase, there are a few basics to keep in mind.

Essential Materials for Coop Construction

Explore different materials commonly used for building chicken coops and their advantages and disadvantages. Some popular options, with their advantages and disadvantages, include (Gray, 2023):

- Wood: Natural and readily available, but requires regular maintenance and protection from moisture and termites.

- Metal: Durable and predator-resistant, but may require insulation to regulate temperature.

- Wire mesh: A highly versatile material and a popular choice to add to the fencing of your run space or protective covering pop or drilled holes. Wire mesh helps prevent predators and pests such as mice, rats, snakes, flies, and other unwelcome guests.

- Chicken wire: A great option to line a run space or provide extra security to a fenced-in yard.

- PVC or plastic: Lightweight and easy to clean, but may not provide sufficient insulation in extreme weather conditions.

Sourcing Materials

If you plan on building your own coop, finding reputable suppliers or repurposed materials to minimize costs and ensure the quality of your coop is essential. However, there are many ways to source your materials, whether brand new or repurposed.

Local Hardware Stores and Building Supply Centers

The best way to ensure you are getting the highest quality materials required for your coop build is to go to a lumber yard, hardware, or building supply store.

In some cases, you may be able to obtain all the hardware, wire mesh, tin roofing, and paint supplies you require, all in one place. You could also get your materials cut to specific lengths at some stores, which is ideal if you are new to building or do not have the tools for cutting your own lumber or metal roofing. What is more, you could ask the experts who work at the store for advice on the best materials and building practices.

The downside to making all your purchases from hardware and building supply stores is that they can be slightly more costly compared to sourcing repurposed materials. However, for some, the convenience and availability of materials may outweigh the difference in cost.

Online Suppliers and Marketplaces

There are several reputable online lumber suppliers and retailers available. As long as immediate delivery is not a top priority for you, this approach can be a more budget-friendly option.

Some online stores offer to cut lumber in precise dimensions; however, not all online suppliers offer this kind of customization, so you may still need wood-cutting tools if you take this route.

Although online lumber shopping might be more budget-friendly, make sure to exercise caution when making online purchases. Not all retailers are reliable and you will also need to consider shipping expenses and delivery timelines. Additionally, certain vendors may stipulate a minimum wood purchase which might not be suitable to your specific needs.

Repurposing Materials

Upcycling from other projects or structures and sourcing from reclamation yards is a great way to recycle lumber that would normally go to waste. Making this building approach eco-friendly, sustainable, and economically viable. A few places to consider looking for materials to repurpose include online marketplaces, yard sales, and reclamation yards.

If you are up for a DIY challenge and want to build a fun and unique coop out of timeworn wooden planks sourced from a century-old barn or leftover wood from someone else's construction project, then reclamation yards are a great resource. These yards maintain a fluid inventory, introducing an element of unpredictability and a sense of discovery to each visit. Lumber is not the only material to look for when selecting repurposed materials. You might even discover nifty odds and ends to make enrichment toys for your birds, like swings or forage boxes. You never know what you might find rummaging through a yard of potential.

However, while these yards possess a whimsy allure, they also come with a downside. It may take some time to obtain all the materials you need since your purchase options are limited to what they currently have in stock and their inventory might not align with your needs.

If you would like to take a shortcut to your DIY coop, consider repurposing and modifying an old shed or other pre-built structure.

Working with reclaimed structures, lumber, or objects can add a touch of individuality to an otherwise standard-looking chicken coop. While gathering all the materials you need might be a little time-consuming, if this notion resonates with you, then upcycling might be a fun venture to consider.

Important Elements to Include in a Coop

A proper chicken coop does not end with building a few walls and a roof. Adding nesting boxes, roots, a secure run space, and comfortable bedding are all part of the coop-building experience.

Nesting Boxes

Nesting boxes serve as a cornerstone for enhancing the flock's general health, productivity, and overall well-being. They establish a special and controlled space for egg laying, ensuring optimal care for both the hens and their invaluable eggs.

Attentively crafted and consistently maintained nesting boxes create a secure and sheltered environment for hens to lay their eggs. This helps prevent the eggs from being exposed to potential damage, breakage, or environmental elements, ensuring that they remain clean and intact. Having designated nesting boxes also provides hens with a sense of privacy and security during the vulnerable act of egg-laying. This reduces stress levels and promotes overall well-being among the flock.

Egg collection is also more convenient and efficient when you have designated nesting boxes. It allows you to easily locate and gather freshly laid eggs without disturbing the hens, reducing stress for both the chickens and the caretaker.

Another significant reason why nesting boxes are important is hygiene. Nesting boxes can be filled with clean bedding materials, such as straw or wood shavings, that help keep the eggs clean and free from dirt or debris. This cleanliness is essential for producing quality eggs and maintaining a hygienic coop environment.

When each hen must have her own nesting box, it helps establish a sense of order in the coop. This helps to prevent competition among hens for nesting space and minimize disturbances that might otherwise arise when multiple hens attempt to use the same area at the same time.

An important note on nesting boxes is that they can encourage broodiness in hens that are inclined to hatch eggs (Arcuri, 2022). A quiet and secluded nesting area provides the optimal conditions for hens to sit on eggs, fostering the natural incubation process.

Optimal Number and Size of Nesting Boxes

As a general guideline, aim to provide around one nesting box for every 3 to 4 hens in your chicken coop (Arcuri, 2022). This ratio helps ensure that there are enough nesting spots for your hens, reducing competition and stress during egg-laying. Keep in mind that not all hens will lay eggs at the same time, so having a few extra nesting boxes can provide flexibility.

For example, if you have 12 hens, having three to four nesting boxes should be sufficient. If you have a larger flock of 24 hens, you might consider having six nesting boxes.

It is important to monitor your hens' behavior and egg-laying habits to determine if you need to make any adjustments. If you notice that hens are waiting in line or showing signs of stress while waiting to use the nesting boxes, consider adding more boxes to accommodate their needs. Providing ample nesting space helps ensure that your hens have a comfortable and stress-free environment for laying eggs.

Suitable Materials for Nesting Boxes

Both straw and wood shavings are commonly used as nesting box bedding for chickens, and each has its advantages and considerations. Choosing between the two will depend on factors such as availability, ease of cleaning, and specific preferences.

Straw provides a soft and cushioned surface for eggs, making it comfortable for hens to lay in, and good insulation, helping to keep eggs at a relatively stable temperature. Additionally, the texture of straw can trigger hens' natural nesting instinct.

However, while straw can encourage nesting behavior, some hens might scratch and peck at it excessively, leading to a messier nesting area. It is also more challenging to clean since it can trap moisture and become soiled quickly, which means it needs to be replaced frequently.

Wood shavings are less likely to become compacted and messy, making them easier to clean. Additionally, wood shavings have absorbent properties, which help to keep the nesting area drier and may even offer some deterrent against mites and parasites. However, shavings provide less cushioning compared to straw.

Whichever option you choose, ensure the nesting boxes are clean, dry, and well-maintained to promote a positive egg-laying experience for your hens.

Roosting Bars

Roosting bars create elevated perches for your chickens to rest and sleep and provide several significant benefits for the health, behavior, and overall well-being of your flock

Chickens have an instinct to perch or roost at night. Roosting bars provide a designated elevated space where chickens can rest comfortably off the ground, mimicking their natural behavior in the wild. Elevated roosts help keep your flock safer during the night when they are more vulnerable, protecting them from potential ground-based predators, such as rodents or snakes.

Additionally, they allow chickens to sleep and rest in an area separate from their nesting boxes. This separation reduces the likelihood of eggs becoming soiled and minimizes the risk of chickens sleeping in nesting boxes. Roosting bars also keep chickens away from their droppings during the night, which promotes a cleaner and healthier coop environment and negates excessive brooding habits.

Roosting is a social behavior that promotes bonding among flock members. Chickens that roost together tend to build stronger social connections and exhibit healthier behavior which can help establish a hierarchy among the flock. Dominant chickens often claim the highest and most central roosting spots, while subordinate ones choose lower positions.

Providing roosting bars with smooth surfaces and rounded edges helps reduce the risk of foot issues, like bumblefoot, which can arise from standing on rough or abrasive surfaces (Smith, 2020b).

Ideal Height

Ideally, a roost should be placed at least 18 inches from the ground and should never be placed over any surface you do not want to get soiled by droppings (Smith, 2020c). Remember that chickens will poop on their roosts, so you should not stack roosts directly above or below one another, above nesting boxes, or feeders. A great tip is to place shallow bins below your roosts for easier cleanup and convenient composting.

Choosing Suitable Materials for Roosting Bars

When you incorporate well-designed roosting bars into your coop, you provide a safe, comfortable, and natural resting space for happy, healthy chickens.

Make sure that you design roosting bars that are appropriately sized and spaced to accommodate your chicken breeds. A general guideline is to allow 10 inches of roosting space per chicken (Smith, 2020c). The bars should be just the right size for your chickens to comfortably grip it with their feet and strong enough to hold the weight of your chickens—approximately 2 inches with rounded edges.

Suitable materials include sturdy natural branches, thick wooden dowels, or PVC piping.

Flooring and Bedding

When selecting bedding and floor litter, consider factors like absorbency, insulation, odor control, and ease of cleaning. Remember that you will regularly have to clean out soiled bedding to maintain a healthy coop environment. The combination of flooring and bedding should promote cleanliness, comfort, and easy maintenance in your chicken coop; and regularly monitoring the coop's condition and the health of your flock will help you adjust your choices to best suit their needs. Some popular choices include straw or hay, pine wood shavings, and, and deep litter.

Easy-To-Clean Materials

Sand, straw, or pine wood shavings are often used since they are cheap and easy to muck out.

Materials such as pine wood shavings and straw can easily be tossed in a wheelbarrow or bucket and tossed on a compost pile. Straw and hay are also great insulators and can help provide your chickens with a little extra comfort, especially during winter months. These materials are also cost-effective and easy to come by.

However, if you decide to use these, you will have to regularly clean and muck out the coop, and replace the bedding with fresh substrate. Otherwise, the risk of developing excess moisture and harmful bacteria and parasites increases.

Sand is one of the most hygienic and easy-to-clean flooring (Mormino, 2013b).

Deep Litter Method

The deep litter method involves layering bedding—typically straw or pine wood shavings—and allowing droppings to accumulate. Over time, the litter breaks down and creates compost, providing insulation and reducing odors. This method focuses on efficient waste management and odor control can be a great time saver for very busy keepers using straw or wood shavings. However, if not executed properly, deep litter can be a health hazard for your chickens, so, before opting for this method in your coop, first try to fully understand how to manage deep litter.

In essence, the deep litter method is garden composting inside your chicken coop. The chickens help with the composting process by regularly turning and scratching the compost pile, providing necessary aeration to the pile.

There are four deep litter requirements: carbon-based litter—such as straw, grass clippings, dried leaves, or pine shavings—aeration, proper coop ventilation, and controlled moisture balance (Kathy Shea Mormino, n.d.).

Carbon-based litter is essential for breaking down and fermenting the nitrogen-rich chicken droppings. Your chickens will naturally provide plenty of aeration to the litter pile; however, in areas that they miss, you will have to turn it yourself to avoid caking. If your chickens are not inside the coop very often you will likely have to do the bulk of the litter aerating.

Make sure your coop has plenty of cross ventilation through open eaves. This can be tricky because you do not want a drafty coop. However, the composting processes produce excess ammonia gasses and moisture build-up needs to exit the coop, so you will need to create a delicate ventilation balance

Maintaining healthy moisture levels is a key part of this waste management system. If too much moisture builds up, your coop becomes the perfect incubator for pathogens that lead to very sick chickens. Spreading fresh litter in your coop regularly, especially on top of water spills, can help manage the moisture levels of your coop compost. A good indicator that your litter has the correct amount of moisture is if it crumbles when you stir it and briefly hold its shape before crumbling when squeezed in your hand. If any water can be squeezed out your litter is too wet, and if it does not hold its shape at all, it is too dry (Kathy Shea Mormino, n.d.).

The best time to start this process is during the spring since it takes months for compost to properly process. Begin by laying down approximately four to six inches of fresh pine shavings. Avoid beginning this process with hay or grass since larger litter will take longer to break down. After your initial four to six-inch litter layer, you can add alternate materials. Make sure that you maintain four to six inches of litter at all times. As your compost breaks down your litter will decrease in depth.

Be sure to constantly aerate your compost. If you spot any caked areas, break these up with a pair of gloves. If you see any areas that are ashen or whitish in appearance, it means that your compost is not being aerated properly so take some extra time to turn those places.

When your litter reaches approximately 12 inches in depth, it is time to remove some from the coop. Make sure to leave at least four to six inches on the floor and restart the process all over again (Mormino, 2013b).

Ventilation and Windows

Making vents for your coop is a relatively simple but essential task. Ventilation and windows allow for temperature regulation, and release of gasses such as carbon dioxide, ammonia, and odor-causing gasses.

Ventilation Openings

Ventilation openings help with fresh air circulation without creating drafts. Pop holes are essentially large openings in the coop that are great ventilation options for small coops. You can cover these with mesh to deter predators; however, the downside to a pop hole is that it does not have a closing mechanism for colder winter months. A way to work around this is to cover these openings with sheets or heavy curtains during the winter to protect your flock from cold drafts.

You can also choose to add ventilation holes by drilling holes in the north and south ends of the coop ceiling, then cover the holes with mesh wire.

Turbine ventilation is another popular method that works by pulling trapped gasses and moisture out of the coop. Turbines are placed on the roof and can be relatively easy to install but note that cross-ventilation is still needed even with a turbine.

Windows or Skylights

Windows or skylights help to provide natural light during the day. Having multiple windows is the best way to add a healthy amount of cross-ventilation in any size coop. Windows have adjustable openings and can be opened depending on the season and can be reinforced with screens or metal mesh. Windows and skylights also have the bonus of allowing plenty of sunlight for your hens during the winter.

Tips for Building a Predator-Proof Coop

Predator-proofing is an important aspect of any coop build, especially if you live in areas where there is a high population of natural wildlife. Thankfully there are a few safety measures that you can include in your coop design to make sure that your flock is as protected as possible.

Understanding Common Predators

Make sure to familiarize yourself with any local predators that may pose a threat to your chickens and take necessary precautions. A few common predators may include dogs, cats, raccoons, foxes, bears, coyotes or snakes, hawks, owls, rodents, and other small mammals.

Reinforcing Coop Security

Predators are crafty and if your coop has a weakness, they will surely exploit it once they discover it. A safe coop protects chickens from predators that come from above and below.

A few things you can do to deter unwanted guests include selecting appropriate wire mesh to reinforce your coop, adding secure locks and latches to doors and windows, installing a fence, and installing a full or mesh roof to your run space. Other deterrents include fencing, motion-activated lights, and noises.

Sturdy Construction With Secure Locks and Latches

Keep predators from sneaking in through open windows and doors by adding latches and heavy-duty locks to areas that they can easily access.

Make sure that you properly secure all access areas at night by double checking that latches have closed behind you, and that locks are fully closed.

Some keepers install an automatic door system for convenience. Depending on your lifestyle this might be something to consider, especially if you are not usually home at night.

Burying Hardware Cloth or Wire Mesh Around the Perimeter

When picking out the metal mesh for your coop, be sure to select a mesh that is ⅜ inches in diameter or less. It is recommended that you also bury wire mesh down the perimeter and under the floors of your coop to prevent invaders such as rats, mice, and snakes (Smith, 2020c).

Additionally, make sure any ventilation holes are secured by covering them with mesh.

Predator-Proof Fencing or an Electric Fence System

If you have large predators such as bears, coyotes, or other large predators you may want to consider adding some extra reinforcement to your yard, homestead, or farm by building tall closed fencing or even an electric fence system.

Motion-Activated Lights and Noise Deterrents

Adding motion-sensitive lights is a great way to scare off potential predators such as raccoons, foxes, dogs, and bears. Additionally, motion-activated noise systems may also be helpful if you live in an area with heavy predator traffic.

Interactive Exercise

Design a floor plan for your chicken coop, taking into consideration the size of your flock, the available space, and the desired layout of nesting boxes, roosting bars, and other essential elements. Sketch or use design software to create a visual representation of your coop design. Also, take a look at the blueprint provided at the end of the book, it can be useful as a general guide or starting point to create your own coop layout!

Chapter 4:

Setting Up Chicken Feed and Water Systems

If you want healthy, happy chickens, you will need to provide your flock with proper nutrition and access to clean water. So, let's have a look at the basic dietary needs of your chickens, and how to choose the right feed, set up a water system, and maintain cleanliness and hygiene in their feeding and drinking areas.

Understanding Chickens' Dietary Needs

We will discuss nutritional needs in more detail in Chapter 5, for now, let's concentrate on how to select the appropriate feed for your flock and some nutritional basics. Understanding how to select feed based on the age and purpose of your chickens will help prevent nutrient deficiencies and help your chickens grow strong.

Nutritional Requirements

Learning the different nutrition requirements of your flock members, depending on whether they are layers or broilers, chicks, or adults, will help you keep them in peak health.

Protein

When a laying chicken produces an egg the protein and nutrient loss is equivalent to that of a human giving birth, except your chicken will lay an egg every day. That is a lot of nutrients to make up for! Without replenishing this protein drain, your hens may stop laying altogether or become ill.

Laying hens are not the only ones who need extra protein. Broilers or actively growing young chickens—especially large breeds—also require a high-protein diet to prevent any deformities, lameness, or illness.

Carbohydrates

The majority of your bird's carbohydrates will likely come from something called 'scratch'. Scratch can either be cracked corn or whole grains—which is a much healthier option. These carbohydrates provide your flocks with energy, as well as some essential vitamins and minerals.

Vitamins and Minerals

When you ensure that your feed contains the proper amounts of vitamins and minerals it helps with egg laying, chick development, and keeping every member of your flock in good health.

Age-Specific Dietary Requirements

What is good for the goose is good for the gander does not apply when it comes to chicken feed.

The nutritional needs of chicks, pullets, and adult chickens vary considerably. What is good for one age group or variety of chicken may not be good for another.

Starter Feed

This type of feed is formulated for young chicks to support rapid growth and development. Typically, chicks are given this type of feed for the first few weeks of their lives since it is high in protein content—approximately 20-24%. However, this type of feed should not be used for chicks past six weeks as it can cause liver damage (Smith, 2020a).

Grower Feed

At the age of six weeks, your chicks are officially teenagers, congratulations! It is time to transition their feed to something more suitable for what their age requires. While grower feed has the same protein content as layer feed, grower feed has nutrients tailored to support strong and healthy body development and growth rather than egg production. Once your chicks are pullets and are around 20 weeks old—or if they unexpectedly lay their first egg—it is time to switch their diet to layer feed (Smith, 2020a).

If you have broiler chickens, they will have to be transitioned to finisher feed instead of layer feed.

Layer Feed

Now that your hens are full-grown adults, it is time for them to get layer feed. This diet will ensure that they are getting the essential nutrients and calcium for consistent health and robust egg production.

If you feed your chicks with feed right off the bat, they will not get the nutrients they require to meet their growth needs. Likewise, if you feed your adult hens starter or grower feed, they will not get the correct nutrient ratios for their specific needs and may overdose on protein (Smith, 2020a).

Choosing the Right Feed for Your Flock

Now that you have a basic understanding of what kind of nutrients chickens need for healthy development, it is time to learn about the different forms of feed available. Choosing the right feed can be a daunting task, considering the wide variety on the market and that each brand advertises that they are the best. The key to selecting the best feed for your flock is to take a close look at the nutritional label and to understand the different forms chicken feed can come in.

Types of Chicken Feed

The different types of feed include mash, crumble, and pellets—which of these you use will depend on how you plan on raising your flock.

Mash is an unprocessed version of crumble and pellets. This feed has a texture that is similar to potting soil and is often used for baby chicks but can be used for adult chickens as well, although this is less practical. To prepare this feed for young chicks, combine mash with hot water until it forms a porridge texture, allow it to cool, and feed it to your chicks or flock (Smith, 2020a).

Crumble is made from whole pellets that crumble into smaller pieces and have a softer consistency. Crumble is suitable for all chickens but it may be easier for small breeds and young birds to consume.

Pellets are made by compressing mash feed ingredients. These ingredients are heated and then compacted to form a pellet that can encourage slower consumption and reduce waste.

Ultimately, the best feed option depends on your flock. Try out different feed styles to determine which one your flock prefers, starting with pellets.

Reading Feed Labels

Interpreting the information on feed labels may seem like a daunting task; however, it can easily be demystified. The first thing to learn is that—in the USA—a non-profit organization called The Association of American Feed Control Officials (AAFCO) models what feed ingredients are safe and what information is required on feed labels. According to the AAFCO (n.d.), the following information must be clearly labeled on feed packaging:

- product and brand name

- what species the feed is used for

- quantity statement—how much product is in the packaging

- guaranteed analysis—the amount of nutrients in the feed product

- list of ingredients—in descending order of highest to the lowest amount by weight

- nutritional adequacy statement—including what stage of life the feed is suitable for

- feeding directions

- manufacturer information—including name and address

Protein Content and Sources

Ideally, your chickens' diet should reflect what they would naturally consume if they were free-ranging or in the wild, which includes the ingredients found in their feed.

Where your chickens get their protein matters. While it is a cost-effective source of protein for many chicken feed options on the market, it may be better to avoid labels that source their protein from soy. Not only is it possible for chickens to have soy allergies, but much of the soy used in animal feed is genetically modified (GM or GMO) varieties, which deplete the environment (Brahlek, 2022).

Alternative protein options to soy include

- pea protein

- worms, grubs, and crickets

- brewer's grains

- nuts and seeds

Additional Additives or Supplements

Some additional nutrients to look for in your feed are prebiotics and probiotics, such as Bacillus licheniformis, Bacillus subtilis, and Lactobacillus acidophilus, which provide support to the digestive, immune, and reproductive systems (Brahlek, 2022).

Recommendations for Specific Age Groups or Purposes

Some feed brand recommendations include:

- Feed for chicks: Little Pecks and Purina® Grower Feed For Chicks (Purina Mills, n.d.)

- Laying feed: Layena, Fresh Pecks by Grubbly Farms, Brown's Layer Booster Chicken Feed, and Scratch and Peck Chicken Layer Feed

- Non-laying feed: Roudybush Low Fat Maintenance or Purina Game Bird Maintenance Chow

- Feed for free-range chickens: Manna Pro Chicken Feed (Atkins, 2023)

Setting Up a Chicken Water System

Water sustains life. So providing your flock with consistent access to fresh, clean water, should be a top priority.

There are various options for water systems such as nipple waterers, gravity waterers, chicken cups, and automated systems. Deciding which is best for you and your flock will depend on your specific needs and flock size. The most important aspect of whichever system you choose is that it remains clean and uncontaminated at all times.

Water Requirements

Just like humans, chickens can become dehydrated and even die from a lack of water. If they drink from a water source that is contaminated with mold, parasites, or bacteria, your flock can fall ill and suffer fatalities. And much like when humans share a drinking cup; if a sick chicken shares its water source with others, they can rapidly spread that illness to the rest of the flock.

Chickens are not dummies either. If they see that their water is dirty, they will not drink from it. They will avoid drinking from water sources that are stagnant or contain algae, dirt, droppings, or chemicals such as chlorine. Chickens will also turn down water that is too warm as this will not help them regulate their body temperature.

This is why choosing a system that is easy to sanitize and provides fresh clean water is vital. When choosing a location for your waterers, make sure to space them apart. This prevents overdominant hens from becoming territorial over or hogging a single water station and bullying away lower-level hens.

Daily Water Consumption Considerations

An adult chicken can drink up to a full pint of water every single day and up to two full pints during hot summer months. So if you have a flock of twenty chickens, plan on providing them with 2.2 gallons (8.3 liters) of water every day during temperate weather conditions and 4.4 gallons (16.6 liters) during the summer (Holte, 2022).

Water consumption can vary from bird to bird depending on their size and if they are layers or free-rangers. Large birds will likely drink a bit more than smaller birds, and laying hens will require more hydration than those that are not layers. Free-range hens may drink slightly less water than other chickens because they are consuming moisture from worms, grubs, and assorted plants. Young chicks need to drink nearly twice as much food as they eat.

While all this may sound overwhelming, once you show your chickens where their water stations are and how to use them, they will remember it and drink when they need to.

The Impact of Dehydration on Overall Health and Egg Production

Dehydration is extremely dangerous for your chickens. Water aids in the chicken's digestive system by helping to break down food into life-supporting nutrients, such as electrolytes. Without proper amounts of water, your birds can develop sour crop and will require medical intervention.

Furthermore, water helps to support healthy brain function. When a chicken becomes dehydrated its blood does not circulate properly leading to disorientation and lack of coordination (Lesley, 2021).

Signs of dehydration in chickens include

- lethargy

- heavy panting

- limpness

- convulsions

- diarrhea

- unconsciousness

- disorientation (Lesley, 2021)

Additionally, if your laying hens do not get enough water to replenish their systems, you will notice that your hens will cease egg production.

Watering Options

There are various methods to provide water for your chickens, you can choose to use water troughs, gravity waterers, nipple waterers, and waterer cups.

Water Troughs

Troughs are tricky. An open water trough or basic container may sound like a quick fix for your watering needs. However, these are not at all recommended because they tend to become soiled or contaminated easily.

Ideally, if you choose a trough it should offer drinking ports and consist of an automatic rather than an open system. These automatic systems are wonderfully time-saving, durable, and suitable for bantams as well as larger chickens. Unfortunately, these can carry a hearty price tag, easily get contaminated with debris, and during the cold winter months, the auto-fill system can break.

Gravity-Waterers

These waterers provide your birds with continuous access to fresh water and save you time from bug-a-lugging water to your coop multiple times a day to clean and refresh troughs.

Nipple-Waterers

This system is a fantastic solution to the water contamination conundrum. Nipple attachments can be attached to a large bucket, allowing your hens to drink whenever they need to, without unnecessarily wasting water by splashing or soiling. While this system works very well for many chicken keepers, the downside to nipple waterers is that not all chickens will understand the concept of using the nipples, no matter how much training and effort goes into teaching them (Lesley, 2021).

Chicken Waterer Cups

There are two different types of water cups. The first requires the chicken to peck at a large spoon-like container for the water to fill up. The second does not require pecking since the water will fill up automatically. Chickens will easily catch on to this system and the most success seems to be with the automatic filling cups rather than the cups that require pecking (Lesley, 2021). Unfortunately, the plastic does degrade over time through wear and tear and the cups often freeze up during cold winters, so this system may not be the best option if you live in a climate that is prone to deep freezing.

Placement and Accessibility

Having water in the coop will not help much if your chickens cannot get to it. Make sure they have access to what they need by playing waterer at appropriate heights or offering 'steps' to a gravity waterer for shorter chickens.

Strategic Waterer Placement for Easy Access

Place your waterers so that they are not too far off the ground for your chickens. You will want to make sure that they can reach their waterers, especially if you have bantams.

Additionally, waterers should be placed in an area that is easy for your birds to get to and there should be more than one water source to negate competition over access.

Multiple Waterers to Prevent Overcrowding

You might only need two or three gravity waterers or twelve water nipples for your flock (Lesley, 2021). However, to prevent bullying and overcrowding at water sources, it is always good practice to have more watering options than the required minimum.

Protecting Waterers to Maintain Water Quality

If you are using gravity waterers or rain barrels, make sure to keep water containers out of direct sunlight and prevent exposure to extreme temperatures.

When a water source overheats the chances for bacterial and mold growth increase exponentially. This water has the potential to harbor dangerous waterborne illnesses that can have a negative health impact (Lesley, 2021).

Tips for Maintaining Cleanliness and Hygiene

Feed contamination is a real issue and keepers must be aware of ways to prevent it and take proper precautions to avoid spoilage. Factors that can lead to feed going rancid are high humidity or moisture, pests, and bacterial or mold contaminants.

You can take steps to prevent feed contamination and spoilage by ensuring your feed is kept in sealed and air-tight containers. Make sure to always close your feed containers and check for any signs that rodents might have been trying to gnaw their way in. Avoid scooping feed with wet or damp scoopers or utensils, and when purchasing feed double check the bags for any signs of dampness or mold.

Feeding your chickens with feed that has gone bad may lead to serious illness, so make sure to check your bins regularly for signs of excess moisture.

Water Quality Management

Clean water is important for your flock. Ensure that the water you provide to your chickens is clean and free from contaminants by regularly inspecting and maintaining the cleanliness of the waterers. Pour out stale or dirtied water from gravity waterers and troughs to remove any debris, pests, or other contaminants. Thoroughly clean the waterers according to manufacturer instructions and replace the old water with clean water.

Regular cleaning and sanitizing waterers will help you prevent the spread of disease and other illnesses within your flock. Make sure to clean your waterers with non-toxic and chicken-safe cleaning solutions.

Pest Control

Implement strategies to prevent pests from accessing feed and water sources. Some basic rules include following proper storage of feed to deter rodents and insects. Regularly inspecting and maintaining feeders and waterers to remove debris or pests. As well as storing food in a cool dry area in airtight containers to avoid mold, pests, and spoilage.

Interactive Exercise

Evaluate your current feeding and watering setup. Identify any areas for improvement in terms of cleanliness, accessibility, or efficiency. Make a list of actionable steps to enhance the feeding and watering systems for your flock.

Chapter 5:

Feeding Your Chickens

Just like humans, chickens require a balanced diet to remain healthy and productive. In this chapter, we will go through the basics of chicken nutrition, how to create a balanced diet, tips for feeding chickens organically, and strategies to prevent and handle feeding issues.

Basics of Chicken Nutrition and Feeding

Chickens have specific nutrient requirements and therefore require a highly diverse diet. As omnivores, chickens enjoy a wide variety of foods including animal protein, eggs, bugs, vegetables, fruits, grains, and even a mouse or two on occasion. It may sound odd to hear that chickens enjoy animal protein considering chickens have neither sharp beaks nor teeth to break the meat down. So then how do they do it?

Understanding the Digestive System

Chickens have unique digestive systems that enable them to break down food and extract nutrients without chewing (Nutrena, 2019).

Their beaks have evolved to be the perfect tool for precision hunting as they pick and peck at the ground to catch their tiny, fast-moving prey. Once they take in and swallow their food, it moves from the esophagus and begins its journey through the chicken's digestive system.

The crop is the first part of the digestive system. Located just below the esophagus, the crop serves as a temporary holding container. Over the course of 12 hours, food is slowly softened and incrementally moved to the gizzard.

The gizzard is the second phase of digestion. This muscular organ acts as the chicken's 'teeth'. Using grit, the gizzard grinds food into smaller digestible pieces.

Food then moves from the gizzard to the intestines. The first stop is the small intestines, where nutrient absorption takes place. From there the food is moved to a sack in the lower intestines called the ceca. The ceca utilizes bacteria to break down any undigested food particles before sending off its work to the bird's large intestines.

During this phase, water is removed from the ingestible food before moving to the cloaca. Once in the cloaca, the waste is mixed with urine—which is the white substance you may notice in chicken poop—and the final waste product is expelled from the chicken's vent (Nutrena, 2019).

Feeding Behavior and Habits

As we discussed in Chapter 4, the best diet for your chickens is a natural one. Chickens love to peck, scratch, and forage for their food. So, giving your chickens as much opportunity as possible to forage outdoors and fill up on natural food sources is ideal.

If your chickens do not get to free-range it may be beneficial to provide them with foraging opportunities by hiding treats in a towel or scattering scratch on the floor. Remember, chickens are creatures of instinct and when they live in an environment that is not natural to them, they may become restless or develop aggressive habits.

Chickens will also feed in groups, so if one chicken eats, all of them will. This can be a problem if you have large-breed chickens. Some breeds have been known to excessively overeat free-choice feed, causing digestive issues, obesity, gout, and illness (Hess & Griffler, 2018).

All chickens need grit and natural insoluble materials for digestion, so it is a great idea to provide your chickens with a grit bucket for them to pick on as needed. However, some large-breed birds tend to gorge themselves on grit which will cause serious digestive issues.

When providing free-choice options to your chickens, be aware of the breed's natural tendencies and instinct towards food and grit. This will save you from having to change feed unnecessarily or having to deal with preventable health complications among your flock.

Daily Feed Consumption Based on Age and Breed

Food consumption will vary greatly depending on the lifestyle you provide for chickens and their breed type. The best way to tell how much food your flock will need is to look at the feed label and assess your flock's behavior and physical health.

Assessing feed consumption can be complicated, especially if you have particularly munchy large-breed birds. You may think they are starving because they finish off every pellet and snack you give them; however, in reality, many large breeds do not recognize when they should stop eating. Therefore, you will need to monitor your chickens carefully when deciding if they need extra food. In the beginning, it might be helpful to weigh your chickens to determine if there are any weight fluctuations and if they need more or less feed during the day.

Creating a Balanced Diet for Chickens

To create a balanced diet for your flock, you will need to consider their nutritional requirements, feed rations and formulations, and incorporate natural foods as much as possible.

Nutritional Requirements

Just like any living creature, chickens have their own set of nutritional requirements to stay happy and healthy. These nutrients include protein, grains, greens, insoluble grit, calcium, vitamin A, vitamin D, and clean water (Hess and Griffler, 2018).

Protein

This is a must for overall growth and energy, especially for laying hens and broilers. A great way to add protein to your chickens' diet is to allow them to roam free. This allows them to munch on plenty of tasty worms, grubs, and other insect morsels. Another way to add protein to their diet is to incorporate soybeans and leftover eggs into their daily mix of food.

Grains

One of the best sources of vitamins B, E, phosphorus, and protein is whole grain scratch such as oats or wheat. However, note that scratch should only make up 10% of your flock's total diet (Hess and Griffler, 2018).

Greens

Chickens need their veggies and greens just like we do! Some great options are to let them eat pesticide-free grass, spinach, lettuce, kale, carrot tops, beet greens, dandelions, and more. Fresh greens provide your birds with essential vitamins and minerals such as riboflavin, calcium, vitamin A, and vitamin E.

Insoluble Grit

Insoluble grit can include things like small rocks and pebbles that act like teeth in a chicken's gizzard and aid in digestion. If you have free-range chickens they will find their own grit; however, if your birds live in their coop, you will have to provide them with the necessary grit at least once a month. Make sure that you select grit that is size-appropriate for your chickens. Without grit, your birds will develop serious digestive issues and can develop an impacted crop.

Calcium and Vitamin A

If your birds are getting a healthy feeding of greens and formulated food they should be getting all the calcium and vitamin A they need. However, deficiencies in either of these are something to be aware of, especially in laying hens. If you notice that your hens are laying soft-shelled eggs, add extra calcium to their diets. This can mean adding calcium to their feed, or natural sources like black sunflower seeds.

Vitamin D

Just like humans, chickens get their vitamin D from the sun. If your chickens are not getting enough sunlight or if you live in an area where the weather is typically cloudy, you may need to add Vitamin D supplements or sources such as kelp, to their food. A deficiency in vitamin D may manifest in weak bones and eggshells.

Water

Everyone needs fresh water daily and chickens are no different. Make sure that your chickens always have fresh clean water available to them.

Feed Ratios and Formulations

In Chapter 4 you learned about chickens' dietary needs, based on breed and stage of life—information that can help you make an informed decision about what type of feed they require. Dedicated feed types are quite helpful, for example, if you primarily have laying birds, they should be eating feeds marketed for "layer" birds. But what should you do if you have a flock of both layers and broilers or what if you have a duck and turkey living with your chickens?

The easiest way to solve this is to have separate coops or segregated feeding times. You might also consider a hybrid feed to accommodate a mixed flock. However, if you select a mixed-flock feed, do it carefully and note that you may need to include additional nutrient supplementation.

Make sure to refer back to Chapter 4 to determine how much protein is appropriate for your chickens and remember that carbohydrates from grains should only take up 10% of a chicken's overall diet.

Incorporating Natural Foods

Regardless of what feed or lifestyle your birds have, they should always be given a healthy supplement of fresh natural foods and treats that include everything from leftover salad, meat, eggs, mealworms, sprouts, watermelon, and foraged bugs.

A few great ways to provide your birds with supplemental nutrition include:

- kitchen scraps and vegetable trimmings

- foraging opportunities in the backyard

- growing fodder or sprouts for added nutrition (Hess & Griffler, 2018)

Feeding Chickens Organically

While both organic and conventional types of feed contain similar essential nutrients, there are some notable distinctions between the two. Conventional feed, widely available at feed stores, is often more cost-effective due to its production methods, which may involve the use of pesticides and genetically modified organisms (GMOs).

Organic feed is cultivated without the introduction of toxic or harmful substances like synthetic pesticides, antibiotics, or hormones, making it entirely natural and untreated. It must adhere to much higher standards in terms of growing and procedures and regulations to earn its organic label. This includes that it must be grown by a farmer who has obtained organic certification and who uses organically certified seeds and farmland. Furthermore, the finalized product must be certified free of all foreign substances—which includes GMOs, pesticides, and hormones.

Given these factors, organic chicken feed is often regarded as superior, due to its purity and high quality (Ranson, 2021a).

Benefits of Organic Feeding

While conventional non-organic animal feeds may be a cheaper option, the overall benefits of feeding your chickens organic food outweigh the price tag. This is especially true for individuals who value consuming organic produce. Remember if your chickens do not eat organic feed, then they will not be producing organic egg or meat products.

Some of the benefits of feeding your flock organic food include a healthier flock, better nutrition, better tasting products, and beyond that, it is the more environmentally conscious choice.

Offering your flock feed devoid of chemicals, toxins, or additives, significantly reduces the likelihood of them developing diet-related health problems (Ranson, 2021a). The nutritional content of organic feed is far superior, with higher levels of essential nutrients like Omega-3 fatty acids.

Additionally, due to its natural cultivation and harvesting methods, organic feed has a more robust and superior taste compared to conventional feed. This means that the eggs and meat your chickens produce   will   also   more   likely   have   a   richer   flavor.

The environmental impact of inorganic farming is catastrophic for the environment and local ecosystems. Due to its cultivation practices, organic feed avoids soil, water, and wildlife contamination, contributing to a healthier and more sustainable ecosystem.

Avoiding Pesticides and Chemical Residues in Eggs and Meat

Studies have shown that pesticide residues can be found in varying degrees in produce depending on the variety of chemicals used, the amount of rainfall, wind, storage time, and washing and peeling (Better Health Channel, 2012). However, some pesticides build up in the body's system over time and accumulate in body fat.

This means that conventionally fed chickens may have trace amounts of pesticides in the meat and eggs you harvest from them.

Additionally, the overuse of antibiotics such as those in non-organic animal feeds contributes to an increase in antibiotic-resistant bacterial strains including E. coli and other harmful pathogens (Better Health Channel, 2012).

Promoting Environmental Sustainability and Regenerative Practices

Organic farming plays a significant role in reducing greenhouse gasses. This is because, unlike conventional farming, it does not use synthetic pesticides or fertilizers that are fossil fuel-based. The results are a much lower carbon footprint and 55% less energy waste compared to non-organic farming (Brook, 2022).

Additionally, organic agricultural practices increase the amount of carbon in the soil, resulting in more stable and nutrient-rich farmland. Studies have shown that synthetic pesticides greatly damage the soil invertebrate communities which are responsible for breaking down organic matter and increasing ground carbon content (Brook, 2022).

It has been estimated that if agroecological best management practices and organic farming were followed globally, soils would be able to absorb more carbon than the conventional farming industry could emit over a thousand years (Brook, 2022). This estimate has significant implications for solving the current global warming crisis.

Organic Feed Options

If you are looking to raise your chickens organically, using the right feed should be your first step. Organic feed options include using certified organic crops, locally sourcing organic grains, and making your own organic feed.

Certified Organic Feed

A few commercially available organic chicken feed options include:

- Nature's Best Organic Feed

- Manna Pro Organic Layer Pellets

- Kalmbach Feeds Organic Layer Feed

- Scratch and Peck Organic Layer Feed

- Prairie's Choice Non-GMO Grower (HappyChicken, 2022b)

Locally Sourced Organic Grains

If possible, it is always best to source your organic foods, including scratch grains, locally. By doing this, you can help support your local organic farmers and the local economy as well as promote sustainability by avoiding excessive packaging and relying on carbon-intensive, long-haul transport.

DIY Organic Feed Formulations

There are many benefits to creating your own chicken feed. One of them is having control over your flock's diet. Another is that it allows you the opportunity to make your feed as natural as possible and adjust the ingredients to your flock's unique needs.

The downsides to DIY formulations are that they can be more costly, some ingredients may not be available to you, it can be time-consuming, and if your chickens are picky eaters, you might end up wasting healthy whole grains and feed.

Remember to proceed carefully when making your own organic feed since your chickens will require different nutrients depending on their age, size, and purpose.

A basic organic chicken feed mix will contain:

- corn

- whole wheat

- peas

- oats

- fish meal

- essential vitamins and mineral supplements

The ratios of these ingredients will vary depending on what your chickens require. It is also recommended to provide your chickens with a free choice of kelp—which is filled with vitamins and minerals—and aragonite—a great source of calcium—along with this formula (Winger, 2022b).

Organic Feeding Strategies

If you are worried about the cost of providing your flock with organic feed, fear not! Two great strategies for reducing your feed needs and maintaining flock health are fermentation and nutrient-packed snacks.

Organic chicken feed is not cheap. A great way to help your feed bag go a little further and increase the nutrient value even more is by fermenting the feed. In Chapter 1, we mentioned fermenting as a nutritious and cost-effective strategy to provide your chickens with the food they need. In this section, we will dive deeper into how to proceed with this process.

While we are discussing organic feed fermentation, take note that the process is the same for conventional feed, DIY formulations, and even scratch grains!

Fermenting is much easier if you have a small flock; however, the process is not impossible for large ones but might become tedious.

To begin fermenting you will need to have the following items and information:

- The exact amount your chickens will need to eat per day

- A large sterile food-grade bucket or large container

- Unchlorinated and filtered water

- A clean cotton tea towel or fine muslin fabric

Measure out how much feed your chickens will eat in one day and pour this into the sterilized bucket—you can sterilize your container or bucket by washing it out thoroughly with boiling water and soap—and cover the feed entirely with the water. Let this stand for approximately 30 minutes before checking on it (The Happy Chicken Coop, 2022). If the feed has absorbed the water, you will have to top it off again.

Next, place your bucket in a warm—but not hot—place in your home. Check on the feed for the first few hours of this process to see if you need to top up the water if necessary. Once the feed is no longer absorbing water, you can stop checking on it.

Let the bucket sit undisturbed for 24 hours before checking on it.

Do not get worried if you find a whitish or clear film on the top of the feed, this is completely normal. If you are unsure, take a sniff. If the feed mixture smells like sourdough, then you are doing great! Top up the water if needed and give it a stir.

Once your feed has been fermenting for a total of 48 hours it is ready for you to use. However, depending on your schedule and preference, you can let it ferment for up to four days (The Happy Chicken Coop, 2022).

Once this process is complete, you will have made one day's worth of feed for your flock. If you have a large flock, you will need more buckets.

Some things to watch out for are:

- black mold on the surface

- fuzzy mold

- a rancid or sour smell

If you notice any of these during fermentation, it means your feed has been contaminated. In this case, you will have to toss the mixture out, fully clean the bucket, and try the process again. Never give contaminated feed to your chickens.

Now that you have an idea of how to save money on organic feed, you might be wondering about organic chicken treats and scratch.

The first thing to do is to read snack labels carefully. If a product is organic it will say so on the packaging. You can also make your own snacks or even raise your own mealworms to guarantee they are fully organic and healthy.

Mealworms are a popular chicken treat and are incredibly nutritious. However, you may not always be able to find an organic source for these. If that is the case, and you are not interested in starting a mealworm farm, the next best thing is to understand where your mealworms are sourced from.

In 2014, due to high rates of contamination, the UK banned all imported dried mealworms and made it illegal to feed chickens with mealworm products to avoid the risk of pathogen contamination (The Happy Chicken Coop, 2022). The reason for this is the concern of a recurrence or similar crisis of the Mad Cow Disease outbreak in the 1980s.

Considering the vast majority of dried mealworm products come from China, a country that has more relaxed food quality standards, use your discretion when it comes to finding a trusted brand.

However, while the UK has banned mealworm imports, millions of happy pets and livestock consume imported mealworms—mainly from China—without incident. If food contamination is a concern for you, take some time to find a local source for your mealworms.

We have discussed the importance of scratch in previous sections of this book, below are some fantastic organic brands to give your hens.

- MannaPro- Mixed Grains Scratch

- MannaPro- SevenGrain Ultimate Scratch

- Scratch and Peck- Organic Scratch n' Corn

- Kaytee- Scratch Plus (The Happy Chicken Coop, 2022)

Avoiding GMOs and Synthetic Additives

Organic feed is higher in nutrient value, free of pesticides, and more importantly, it is free of GMOs.

There is a great deal of controversy over genetically modified foods. Unfortunately, science cannot presently determine what the long-term health effects of consuming GMOs may be.

While mega-companies that manufacture genetically modified food items, strongly claim that GMOs are entirely safe, numerous animal studies have shown that GMO consumption may lead to organ damage, immune system malfunctions and deficiencies, accelerated aging, and even infertility (The Happy Chicken Coop, 2022c).

However, more studies are needed to confirm what—if any—health and safety risks GMOs pose to humans. Currently, research is being conducted on the startling rise in allergies, endocrine disruptors—which impact pregnant women, and children under the age of 7—and antibiotic resistance (The Happy Chicken Coop 2022c).

By choosing to use organic chicken feed, you are promoting the flock's health and protecting yourself and your family from any potential harm and potential side effects of GMOs and chemicals found in conventional feed.

Incorporating Natural Supplements

Incorporating natural supplements, such as probiotics and herbs, into your chickens' diet is a great idea, especially if they do not get out of their coop and run space much.

If you are fermenting your feed then you have already added plenty of probiotics to your chickens' diets and adding more is not necessary. Supplements should not make up more than 10% of your flock's daily food source and should be treated merely as a little something extra.

Certain herbs can be great for overall flock health and wellness; however, always consult a veterinary professional before giving anything to your hens, as randomly supplementing certain herbs can do more harm than good.

Some healthy herbs include

- calendula

- oregano

- thyme

- sage

- tarragon

- fennel

- parsley

- marjoram

- basil

- bay leaves

- lemon balm (The Pioneer Chicks, 2023)

Preventing and Handling Feeding Issues

Chickens are special creatures with varying eating habits. While some are prone to overeating, others are overactive and need more calories. Molting and brooding can also affect their eating habits. In the end, your bird's eating habits depend on breed, individual temperament, and hormonal changes.

Overeating and Obesity

Yes, obesity in chickens is a real issue. As mentioned in earlier chapters, some larger breeds of chickens do not know when they have had enough to eat and can binge eat themselves to death.

To prevent obesity and other health complications that certain breed types get from overeating, try implementing these two rules:

- Control feeding and portion size.

- Limit treats and high-calorie foods.

Control the portions you are giving your flock and consider providing feed in the afternoon rather than first thing in the morning for free-ranging chickens. Additionally, only give your chickens snacks or treats after they have had their daily feed ration.

If your birds are already overweight, hold back on some of their higher-calorie snacks and treats.

Feeding Strategies for Specific Situations

There are certain situations when your hens might not eat as much, such as during molting season and when they are brooding.

Molting

Molting happens yearly and your birds may need a little extra food to replenish the energy it takes to molt. Adjusting their feed to support feather regrowth during this time will help keep your chickens healthy. Adding a little extra protein to their daily food rations can help them grow strong and shiny new feathers.

Broody Hens

When a hen is brooding—laying and incubating eggs—she will consume up to 80% less feed than she normally would (Smith, 2020e). Because of this, a brooding hen may become malnourished. If you notice your hen becoming too thin, place a private food and water source closer to her nest. This way the hen can access nutrition without walking too far from her eggs and causing stress.

Managing a Larger Flock

Strategies for Managing and Scaling Up a Chicken Operation

Farming is one of the oldest business ideas in history. If you are looking to scale up your chicken operation and turn it into a profitable business, carefully consider the following questions:

- Do you have enough space to safely and comfortably home more chickens?

- Do you have enough egg-laying breeds? Or would you need to purchase more?

- Do you have the right chicken breeds to suit your business goals—egg layers or broilers?

- What is the local market demand? Do you have enough potential customers to justify expanding your farm/business?

- Can you afford to invest in automated equipment and machinery to make your farming more efficient? Are you able to afford the extra feed and housing costs of keeping more chickens?

- Can you diversify your business? Having multiple streams of income can provide a financial buffer during slow seasons. Diversifying your poultry farm might mean selling chicken manure to local farmers and gardeners or opening a farm stand.

Efficient Feeding, Housing, and Flock Maintenance

In order to expand your farming operation, you will need to prioritize disease prevention, feed costs, and coop space. Investing in automated systems may help save time with watering and feeding; however, you still have to do daily cleaning and maintenance, especially with a larger flock.

Interactive Exercise

Create a customized feeding plan for your flock. Consider their age, breed, and specific nutritional needs. Calculate the appropriate feed quantities and monitor their feed intake to ensure a balanced diet. Track any changes in their health or productivity as a result of the new feeding plan.

Make a Difference With Your Review

Unlock the Power of Generosity

"Kindness is like a pebble dropped in a pond; the ripples can reach farther than you imagine" – Anonymous.

People who share their knowledge create a ripple effect that impacts the world. So, if we've got a chance to do that together, let's give it a shot.

To make that happen, I have a question for you...

Would you share your thoughts to help someone discover the incredible world of raising chickens, even if you never got credit for it?

Who is this someone, you ask? They are just like you, or maybe just like you used to be. Curious, wanting to learn, and in need of guidance but unsure where to find it.

My mission is to make the magic of raising chickens accessible to everyone. Everything I do stems from that mission. And, the only way for me to accomplish that mission is by reaching... well... everyone.

This is where you come in. Most people do, indeed, judge a book by its cover (and its reviews). So, here's my ask on behalf of a fellow chicken enthusiast you've never met:

Please help that budding chicken keeper by leaving a review for *Raising Chickens* by Claire Hennington.

Your gift costs no money and takes less than 60 seconds to make real, but it can change a fellow chicken lover's life forever. Your review could help...

...one more family sustain their garden ecosystem.

...one more person enjoy homegrown eggs.

...one more aspiring farmer start their own flock.

...one more backyard become a haven for happy, healthy chickens.

To get that 'feel good' feeling and help this person for real, all you have to do is, and it takes less than 60 seconds, leave a review.

Simply scan the QR code below to leave your review:

[Review QR Code](insert_qr_code_here)]

If you feel good about helping a faceless chicken enthusiast, you are my kind of person. Welcome to the club. You're one of us.

I'm that much more excited to help you raise your chickens happier/healthier/easier than you can possibly imagine. You'll love the insights into the chicken world that I share in the coming pages.

Thank you from the bottom of my heart. Now, back to our regularly scheduled egg-citing adventure.

– Your biggest fan, Claire Hennington

PS—Fun fact: If you share something valuable with another person, it makes you more valuable to them. If you believe this book will help another chicken enthusiast, send it their way. Let's create a ripple effect of chicken love!

Chapter 6:

Chicken Health and Wellness

Beyond making sure your flock has a healthy nutritious diet, you will also have to keep an eye on their overall health and wellness. In this chapter, you will learn about common health issues in chickens, how to perform regular health checks, tips for maintaining a healthy flock, and strategies for dealing with chicken diseases and pests as well as properly restraining a chicken for health check-ups and injury.

Common Health Issues and Treatments

One of the difficult aspects of keeping chickens is having a sick bird. Unfortunately, the treatment options for bacterial or viral infections are limited and their presence may ultimately end in saying goodbye to a favorite bird. Ailments like parasitic infections are thankfully a bit easier to take care of, especially when they are external. However, while birds are resilient, they are also very fragile when their health is compromised; therefore, to avoid any unnecessary grief, it is best to be vigilant and regularly check your chickens for signs of pests or illness.

If you do not have an avian vet or one familiar with chickens, there are some steps you can take at home to identify and treat certain common infections.

In most cases, treating your chicken for illnesses or injuries will require some basic first aid and a little extra loving care, which includes isolation, providing them with safe shelter, keeping wounds clean, and providing food and water.

However, other illnesses may require veterinary or professional assistance, since you will need a specialist who is capable of collecting specimen samples, prescribing medication, implementing advanced care, and or biosecurity measures. If you can, always contact a specialist if you notice any signs that a chicken may be unwell.

Common signs that your chicken may be sick are

- lack of appetite

- watery or bubbly eyes or nose

- wheezing or rasping when breathing

- frequent sneezing or coughing

- puffy or swollen looking around the eyes

- sudden change in your chicken's natural odor

- change in stool consistency

- smell, or visible blood, or parasites in droppings

- enlarged crop

- balding spots or loss of natural color pattern

- unkempt or brittle feathers

- hunching over

- straining to pass an egg

- drooping wings

- self-isolating

- inflamed, swollen, or the appearance of sores on comb or wattles

As mentioned earlier, the moment you notice something wrong with one of your birds, follow best practices and remove the sick chicken from the flock immediately.

While illness is not common in well-maintained chickens, there are some things your chickens may come in contact with that can make them sick, such as ingesting blue-green algae, cedar wood, lead, and certain molds, or ingesting processed human food like junk food, pesticides, herbicides, rodenticides, Polytetrafluoroethylene (PTFE), too much salt, alcohol, avocados, and chocolate (Barns, 2019).

Respiratory Issues

Chickens are also vulnerable to various diseases that target their respiratory system. These contagions can have viral, fungal, bacterial, or even mycoplasma (bacteria that lack a cell wall) origins and may also impact the nervous system. These infections may include the following (Jacob, n.d.):

- Viral infections: Avian Influenza, Fowl Pox, Infections Bronchitis, Laryngotracheitis, and Newcastle Disease

- Bacterial infections: Infectious Coryza, Fowl Cholera, Colibacillosis (E. coli), Bordetellosis, Salmonella, and Bumblefoot

- Mycoplasma infections: Mycoplasma gallisepticum, Mycoplasma synoviae, Mycoplasma meleagridis

- Fungal Infections: Aspergillosis

Signs of Respiratory Infections

Respiratory infections in chickens may cause any variation of the following symptoms:

- coughing or sneezing

- gasping, open-mouthed breathing, or difficulty breathing

- wheezing, rasping, or rales

- discharge from eyes and/or nose or conjunctivitis

- swollen eyes

- swelling of wattles and/or face

- shaking or twisting of head and neck

- face discoloration or darkened head

- a bluish-purple look

- warts or scabs

- paralysis or prostration

- lameness

- lethargy or failure to thrive

- reduced egg production or laying thin-shelled eggs

- diarrhea that may be discolored

- stunted growth (Jacob, 2023)

Treating Respiratory Infections

Antibiotics are not effective against and do not cure viral respiratory infections in chickens. Without veterinary assistance, there is no way to know if your bird can be cured or if the pathogen they are carrying is dangerous. Do not return a chicken with a respiratory infection to the flock until you are absolutely certain that the bird is no longer ill. Even if the bird does not show signs of illness for a few days, they may still be carriers for the infection.

Unfortunately, while the majority of bacterial diseases that your chickens may contract are treatable with antibiotics, they are not curable. Antibiotics only help to manage the symptoms. Returning the sick bird to the flock makes them carriers of the disease, which means they can spread illness to the rest of your flock. Since bacterial infections cannot be cured, chicken keepers only have a few options available to them.

Depending on the bacterial pathogen, flock owners may choose to cull the infected chicken(s) and do their best to eradicate all signs of the disease from the rest of the flock. This includes fully sanitizing the coop environment, testing other flock members, and choosing to depopulate the flock based on positive pathogen results.

Another option is to reintroduce the disease carrier to the flock and consistently treat the entire flock with antibiotics, fully accepting that your entire flock will always be infected and will be disease carriers to other healthy birds. If this is the option you choose, it is highly recommended not to introduce new chicken members to the flock or allow the flock to free range where they may interact with other fowl from neighboring farms or small flocks (Cruz-Rincon, 2023). Do not allow new chicks to hatch from the infected flock, as many bacterial infections can cause serious birth defects and can cause chicks to hatch already infected. Your chickens must be kept as a closed flock and allowed to die off naturally to prevent the spread of disease. Once this occurs you can fully sanitize the coop and start again with a new healthy flock.

The option you decide on will depend on the type of bacterial infection your birds have and what you and your veterinarian agree on as a suitable solution. Furthermore, depending on the disease, you may be required to inform anyone who comes into contact with your chickens or their eggs and/or meat that your flock is infected (Cruz-Rincon, 2023).

Fungal infections can be fairly straightforward. If the infection is severe, antifungal medication may be necessary—especially if the infection is respiratory—but for the most part, treating your bird(s) with good probiotics can help them fight off the fungal infection naturally. Make sure to continue practicing biosecurity measures and provide extra ventilation and fresh air for your infected birds.

Thankfully, there are vaccinations available that can protect your flock from many viral and bacterial infections. Make sure to talk to your veterinarian about your options when starting a new flock and find out if your hatchery vaccinates its chicks against any common pathogens.

Parasites and Pests

Everyone has parasites, even you! But not all parasites are good and unfortunately, chickens are fairly prone to having them. Parasites can include external ones like fleas, mites, ticks, and fleas, or internal ones like coccidia, and other worms.

In most cases, pests and parasites will merely cause minor irritation. However, more severe cases can result in death. Parasites may also cause a reduction in egg-laying, lethargy, changes in appetite, or unhealthy stools.

Conducting regular health check-ups—such as checking through the feathers of your chickens for pests like ticks and mites and inspecting their poop—can help in early detection of parasitic infections and pest infestation.

A few things to know before treating a parasitic issue is that, unlike viral or bacterial diseases, parasites have lifecycle patterns and can be passed from bird to bird—especially wild birds that come into contact with your chickens—and quarantining a bird with parasites will likely not stop the spread of infection to the rest of the coop. It is best to assume that if one chicken has parasites, all of them do.

External Parasites

Symptoms of external parasites, like mites, lice, and fleas, include itching, excessive or aggressive preening, weight loss, broken feathers, bald spots, reduction in egg laying, vitamin deficiencies, and paling coloration of combs and wattles.

Mites are very common among chickens. One of the best ways to prevent your hens from becoming unwilling hosts is to provide them with the space to have regular dust baths. Unfortunately, this does not always keep the mites away, therefore it is wise to check your chickens and provide treatment as needed.

There are several types of mites, including the Northern Fowl, Red Chicken, and Scaly Leg Mites. A good indicator that your bird has a mite problem and may need intervention is if you see it excessively preening or pecking at its skin, or notice sudden bald spots outside of the molting season. You can help your chicken get through this rough patch by adding extra nutrients and protein to their diet and identifying which mites are causing the problem.

For northern fowl mites, a generous dusting with wood ash on your chickens and the entire coop can be very effective in eradicating them. Red chicken mites on the other hand typically feed and bite their hosts at night when they are roosting. If you notice your hens avoiding their roost at night, it is a good indicator of a red mite infestation. Unfortunately, these parasites are notoriously difficult to exterminate and the best way to eradicate the infestation is to re-home your flock and heavily treat the coop against mites for a minimum of six weeks. In extreme cases, you will unfortunately have to burn and remove the red mite-infested coop.

Thankfully, scaly leg mites are relatively easy to eradicate, although leaving an infection untreated can lead to lameness or death. These mites burrow under the scales of a chicken's legs, which leads to pain and extreme discomfort. To get rid of these mites, soak the infected chicken's legs in warm water to soften the skin. Be sure not to pull off the scales on your bird's skin, but gently remove any loose skin. After soaking and cleaning its legs generously with vegetable or olive oil, gently brush them with a toothbrush to ensure the undersides of the scales are fully covered. Gently wipe off any excess oil and then generously cover its legs with petroleum jelly, such as Vaseline. The Vaseline will suffocate the adult mites as well as any eggs that were laid under your chickens' scales (The Happy Chicken Coop, 2022).

Chicken fleas and lice are common parasites among chickens. Fleas are typically brown and are large enough to spot between or on your bird's feathers, whereas spotting lice might be a little bit trickier. Lice and fleas tend to be more ferocious during the summer months so it is advised to stay on top of any sign of these pests.

The best way to get rid of these pesky creatures is to dust all your chickens—even the ones without fleas—with poultry-safe diatomaceous earth. After thoroughly dusting your birds, clean your coop by removing all bedding from nesting boxes and any flooring such as hay or pine chips and dust the coop top to bottom with the diatomaceous earth. Make sure to get into all the nooks and crannies of the coop including the roosts.

Repeat this process again two weeks later and it should free you and your birds from any fleas or lice. If the second treatment was not enough, you can repeat the dusting process once more two weeks after that. Before using diatomaceous earth on your birds, make sure to consult a veterinarian as there is some controversy over whether diatomaceous earth is entirely healthy for birds (The Happy Chicken Coop, 2022). So, be aware that although this is a common flea treatment for many chicken keepers and other pet owners alike, you will be using it at your own risk.

Internal Parasites

Internal parasites fall into two groups, worms or protozoa—coccidia, being the most common among chicken flocks. Worms and coccidia are fairly common in both free-range and coop-confined flocks. Just like in human bodies, low levels of intestinal parasites do not typically cause any issues, but if the parasites begin to multiply beyond what the immune system can handle, the infection becomes a problem. Internal parasites in chickens can result in anemia, weight loss, diarrhea, low-quality egg production, failure to thrive, and a higher susceptibility to other infections.

Worms can be passed between chickens through fecal matter, contaminated water, or food sources. Aside from physical observations in behavior, a good way to tell if your chicken has a severe worm infection is if you spot them in your chicken's poop or—if a large roundworm made it into a hen's reproductive tract—encased in a newly laid egg.

Worms can be treated by following strict sanitation practices including fully cleaning and sanitizing the coop, nesting boxes, waterers, and feeding troughs. In addition to this, you will need to contact a veterinarian for anti-parasitic medication to treat your entire flock. Do not introduce new birds to your flock until your worm problem is taken care of.

Unlike worms, coccidia is a unique internal parasite in that they are "host-specific" and will not affect any other livestock, including turkeys. Coccidia is almost always found in low levels among chicken flocks, and because of this, most chickens will develop immunity to this parasite over time (University of New Hampshire, 2023). Coccidia causes intestinal tissue damage in the infected chicken, meaning you may notice bloody or watery stools, dehydration, anemia, and lethargy in a chicken with a severe coccidiosis infection.

It is not possible to eliminate this parasite, so to control it, maintain a clean and properly sanitized environment and ask your veterinarian for medicated food recommendations.

Implementing Regular Parasite Prevention Measures

Keeping intestinal parasites such as coccidia away is relatively straightforward. Cleaning water and food feeders daily, maintaining proper nutrition, cleaning your coop regularly, and vaccinating your chickens for coccidia early on, may help your flock from developing serious symptoms (University of New Hampshire, 2023).

Nutritional Deficiencies

When your chickens do not get the nutrition they need it can take a big toll on their health. Your flock will likely have different nutritional requirements depending on their stage of life. Because of this, it is important to recognize the signs of nutritional deficiencies in your chickens. Ideally, you should be able to prevent any deficiencies in all of your birds by providing appropriate dietary supplements and food resources.

Symptoms of Vitamin and Mineral Deficiencies

A few signs of nutritional deficiencies are brittle feathers, stunted growth, thin or soft-shelled eggs, anemia, and birds prone to injury, illness, and parasites.

If your chickens are nutrient deficient, the first signs you will typically notice are due to a lack of vitamin B. These initial symptoms may include reduced egg production, lethargy, head tremors, digestive issues, decrease in heart and respiratory rate, unhatched eggs, slow growth, diarrhea, muscle atrophy, loss of appetite, and weight loss. In adult birds, these symptoms may appear within 2-3 weeks of birds experiencing a deficiency (Korver, 2023).

Supplementation Strategies to Address Specific Deficiencies

Typically, vitamin deficiencies occur because you are neglecting to add a complete vitamin and mineral premix in your chickens' diet (Korver, 2023). At the first sign of nutrient deficiency, immediately add balanced and complete nutrient supplements to your chickens' diet.

Egg-Laying Issues

Your chickens may have difficulties or stop laying eggs for a variety of reasons, including old age, change in weather, natural light, molting season, stress, illness, or nutritional deficiencies. Understanding the common issues that can affect egg production will better prepare you to differentiate between when a hen stops laying for normal natural reasons or due to an underlying condition.

Before worrying about why your hen is not laying, double-check around the coop to make sure that she has not made an alternate nest somewhere. Sometimes, if chickens are not comfortable in their nesting boxes they will improvise, something that is especially true of free-ranging hens.

Next, ask yourself if your chickens are getting enough daylight. Has the season changed or has the weather been especially dark and rainy? Perhaps your chickens just need some help from an artificial light source.

Another element to consider is that your hen might be under stress. Assess your coop and flock for any signs of pests or small predators that might be entering and disrupting your birds. Ensure that your coop is not overcrowded and that your hen is not being bullied.

Molting is another reason for chickens to take a little break from egg laying and typically occurs every year in the fall. During this process, chickens will lose old feathers and regrow new ones. Molting takes a large toll on a chicken's energy levels; so add a little additional protein to your chicken's diet during this time, until they have regrown all their new feathers.

Finally, assess the overall health of your birds and their diet. Hens will stop laying entirely or produce unhealthy eggs when they are nutrient-deficient. Surprisingly, giving your flock too many treats or over-supplementing vegetable scraps in your bird's diet can cause them to prefer filling up on those tasty morsels and ignore their nutrient-dense feed. This can cause a depletion in calcium, protein, and other essentials resulting in your chicken laying fewer eggs (Briggs, n.d.).

Egg Binding and Prolapse

Egg binding is a serious condition that can be life-threatening if not treated swiftly; thankfully, it is not very common. Egg binding occurs when an egg becomes stuck inside of a hen's reproductive tract and she is unable to pass it. During normal laying, the egg travels to the chicken's uterus with its pointy end down. Next, the protective shell develops over the egg and it begins to rotate so that the hen passes the rounded bottom of the egg first. If egg binding occurs the egg is not able to rotate properly and becomes stuck. One possible cause for egg binding is a lack of calcium in your hen's diet, so make sure to monitor your flock's calcium intake (Lesley, 2020a).

Symptoms of egg binding can look different from chicken to chicken and may include:

- Waddling, which is sometimes referred to as a 'penguin walk'.

- Tail pumping is a tell-tale sign that your hen is egg-bound. It may look like she is trying to twerk by squatting low and pumping her tail up and down. This action sometimes helps the hen to pass her egg; however, this is not always successful.

- A loss in appetite

- Lethargy, exhaustion, or a general uncomfortable and depressive demeanor.

- Shaking may also result from excessive straining to pass an egg.

- Not pooping or having very watery or foul-smelling poop. This typically occurs because the bound egg causes enough pressure to close off the chicken's intestinal waste vent.

If you notice that your chicken is not able to pass her egg there are a few things you can try before rushing to the vet. Carefully secure your hen so that she cannot flap around or escape, and perform an internal exam. With a latex-free exam glove generously covered in Vaseline, gently insert your index finger into your hen's reproductive vent. Orient your finger so that it is pointed forward and slightly upward, you should be able to feel the bound egg with your finger inserted 2 inches or less into your hen's vent. If your finger goes past 2 inches and you still do not feel an egg, then your chicken is not egg bound and there is something else wrong with her (Lesley, 2020a).

Once you have confirmed that your hen is egg-bound, fill a bucket or small tub with a ratio of half a gallon of warm water and half a cup of Epsom salts. Make sure that the water is not too hot, it should be just slightly warmer than lukewarm. Gently immerse your hen in the water up to her abdomen so that her reproductive vent is fully submerged. Soak the hen for about 20 minutes before taking her out and drying her off. Give her a quiet, secluded, dark area away from the coop where she can peacefully attempt to pass the egg. Repeat this process after a few hours if she has not passed the egg. If after three soaks your hen still has not passed the egg, make a call to a veterinarian for advanced treatment.

Be extremely gentle with your chicken during this process, do not force her to submit. The egg can break inside of her body resulting in extremely serious consequences. Therefore, do not attempt this treatment unless you are comfortable and experienced with safely restraining your chickens.

Prolapse occurs when the hen's oviduct—the part of the reproductive tract that pushes the egg out—does not properly retract and instead remains exposed outside of the vent after an egg is laid. When this happens, blood flow is cut off to the oviduct, which can result in death. If this happens to your chicken, call a veterinarian as soon as possible.

You can tell that your hen has prolapsed when you notice that she or any other chickens are pecking at her backside. Take a look at your chicken's vent, you should see something resembling a severe hemorrhoid. The prolapse may be bloody, discolored, and/or very swollen (Lera, 2022).

Soft-Shelled or Shell-Less Eggs

Decreased egg production, egg binding, and prolapse are not the only effects a lack of nutrients has on egg laying. Nutrient-deficient hens can also be prone to laying soft-shelled or shell-less eggs.

Oddly enough, soft-shelled and shell-less eggs are a more common occurrence in the summer months. This is because when chickens get hot, they tend to pant to help decrease their body temperature through evaporative cooling, which ends up depleting their calcium (Lesley, 2020). In order to prevent defective shells and egg binding, add extra calcium to your hens' diet in the summer if you notice them panting frequently or if you spot any shells that are not up to snuff.

Regular Health Checks and How to Perform Them

Please note, during a case of emergency—such as injury or illness—an at-home physical examination is NOT an adequate substitute for professional veterinary care.

You should only perform an examination yourself for preventative measures, such as checking for pests, signs of illness, or injuries, and performing emergency first aid. Even after performing a physical examination, if your chicken is behaving abnormally, you should immediately quarantine the bird and call a veterinarian.

Understanding the Importance of Regular Health Checks

By now it should be clear to you how important regular health checks are in preventing and managing health issues in chickens. To properly check your chickens you must know exactly how to perform the basics such as safely restraining a chicken, handling their wings, being cautious of their talons, and how to perform basic first aid for injuries. To do this, you will also have to understand the basics of chicken anatomy and how to spot early warning signs.

Detecting Early Signs of Illness or Injury

Chickens are vivacious, adventurous, and active creatures. While these traits are some of their best qualities, they can also be some of their greatest weaknesses. Chickens can fly into unstable objects, knock over feed and waterers, get caught in chicken wire, and even injure themselves—or other chickens—during a fight.

We previously covered various signs and symptoms of illness; unfortunately, many of these symptoms overlap with those of injury which makes determining the cause of a bird's discomfort difficult to discern without a visible external injury.

Signs of injury include:

- signs of bleeding or dried blood

- cuts or sores

- flailing or erratic flapping

- limping

- pale comb and/or wattles

- changes in posture

- wing drooping

- reduction in egg-laying

- lethargy

- loss of appetite

- inactivity

- isolating or hiding (Mormino, 2016)

If you notice any of the above-mentioned signs that your bird may be injured—or ill—take immediate action and perform a physical examination. If you do not perceive any obvious signs of external injury, isolate your bird in a comfortable quiet location, call your vet, and monitor the chicken for signs of deterioration or sickness.

Monitoring Overall Flock Health and Productivity

Keeping a chicken log is a great way to keep track of their overall health, productivity, and behavior over time. It can especially help you identify changes in behavior that you might have missed otherwise.

Keep an eye out for any symptoms, or odd or sudden behavioral changes. Remember that depending on the problem, chickens can hide their illness or injuries very well, so you may not take notice at first—unless you have a written record of that bird's behavioral patterns.

Performing a Physical Examination

In order to develop the skills to perform a basic physical examination on your chickens, you first need to learn how to properly restrain your chicken and when it is necessary to do so for both your safety and that of your bird.

Before you can restrain a chicken, you need to know how to catch it. If you have a docile chicken, then catching and picking it up might not be much of an issue; however, if you have a skittish bird, you might be in for a workout. To avoid chasing down an injured chicken and potentially causing more harm or further injury, you can employ a few alternative methods to catch your flighty friend. Such as bribing it with treats, using a chicken catching net (only use this method if the bird is uninjured), catching it in its nesting box, or using other distractions.

Once you have successfully captured your bird, swiftly and carefully restrain it. One method you can use is to gently but firmly pick your chicken up with one hand over its wings from the top and your other hand supporting its legs and breast. You may also choose to grab your chicken with two hands and then maneuver the bird to have one wing pressed carefully but securely to your abdomen, while your one hand holds the chicken's wing and your other hand supports its legs and breast, or simply tuck the chicken under your arm. Make sure its wings are secure and that it cannot flap around. If your chicken is not calming down—if it is still pecking, scratching, or squirming—after securing it, place a lightweight towel, dark T-shirt, or an avian blindfold over its head (European Commission, 2018). Being enclosed in darkness should calm your chicken and stop it from moving around.

Be careful not to squeeze your bird too tightly. If you notice any signs of shallow breathing or sudden lethargy, you may be impeding its ability to breathe. Your hold should be gentle but firm enough to keep your bird's wings from flapping around and causing injury to you or it.

Please note, if you suspect your chicken has a broken wing, then gently attempt to restrain it and keep its wing in a natural position, swaddle it in a towel so that it cannot flap, and take it to the nearest veterinarian (Wills and Ludlow, 2019).

After you have successfully restrained your bird, you may proceed to inspect its overall physical health. This may include checking for a bound egg, broken wing, or leg, listening for wheezing, or simply checking the weight and overall physique of your chicken.

Checking Body Condition and Weight

Once you have securely restrained your chicken, you can check for any changes in weight by feeling around its body for areas that usually carry the most weight. If you notice any signs of weight loss such as protruding bones or lack of muscle tone, this may be an indication that something is going on with your bird.

Inspecting Eyes, Beak, and Comb for Abnormalities

Begin at your chicken's head and work your way down (Sato & Wakenell, 2022).

- Comb: A healthy comb should be red, warm, and turgid, and there should be no indication of abnormalities such as scabbing or growths.

- Eyes: A healthy chicken should have its eyes wide open and should look bright and clear with a visible pupil and colored iris—which may be gray or bluish in chicks and copper-red in adults. Its eyelids should not be swollen, red, or drooping over the eyes. You should not see any kind of discharge or cloudiness.

- Beak: A healthy beak should be smooth and pointed at the end. There should not be any cracks, breaks, or deformities. The nares (nostrils) should be clear and free of any discharge, mucus, scratches, or crusting.

Next, notice how your chicken is holding its neck. It should be able to hold its head up and have healthy muscle response and tone.

Examining Feathers, Skin, and Feet for Signs of Parasites or Injuries

Take a look through your bird's feathers for any signs of pests or parasites and take note of any bald spots or feather loss. Ask yourself if it looks like the feathers have been plucked out or if your chicken is just molting. Look for signs that your bird's feathers are brittle or messy. Finally, check its bottom and make sure the feathers surrounding its poop vent are clean. There should be no signs of fecal matter—this may indicate that your chicken had diarrhea—swelling, blood, or scabs around the vent. If you see any blood or scabs, this may be a sign of pecking and indicative of a larger issue.

Next, inspect your chicken's legs and feet. Healthy scales are smooth and close together. Its feet should be free of abrasions, swelling, scabbing, lesions, or ulcers. If you notice any lifting, crusting, or flaking, on your bird's legs, they might be infected with scaly leg mites (Sato and Wakenell, 2022).

Monitoring Behavior and Flock Dynamics

Monitoring flock behavior helps you to address dangerous behavioral issues and spot early warning signs of illness, injury, or pests. When you record baseline behavior and flock dynamics—such as pecking order—you can more easily spot any unusual changes and better accommodate your needs.

You might, for example, notice that a usually spirited chicken is suddenly getting bullied or that there is a sudden increase in aggression or restlessness in your flock as a whole.

Once you notice these changes, you can assess for any signs of parasites, predators, or sickness and take the necessary action. Remember to always act immediately at the first sign of illness to give your bird the best possible care and prevent anything from spreading through your flock.

Understanding the individual personalities of your chickens as well as flock dynamics also helps to determine if you have enough coop and run space, enrichment (such as forage toys and swings), food and water accessibility, roost space, or nesting boxes, for the size of your flock.

Changes in Eating or Drinking Habits

Aside from illness or injury, there are a few reasons a bird might not be eating or drinking enough, including:

- brooding

- old age

- being low on the pecking order

- not having enough food available

- favoring food sources other than feed

However, if you notice your bird is drinking or eating excessively this may be a sign of trouble, such as a parasitic infection.

Social Dynamics and Behavior

Other changes to monitor and make a record of include:

- lethargy

- aggression

- unusual behavioral changes

- signs of bullying

- stress

- aversion to roosting or nesting

- odd walking or twitching movements

- isolation

- brooding

- fighting over food or water

- avoidance of spaces or corners of the coop

- fright or sudden commotion (Freedom Ranger Hatchery, n.d.)

Tips for Maintaining a Healthy Flock

Biosecurity Practices

We have touched on biosecurity measures in previous chapters; however, in this section, we discuss this process in detail, including the details and gravity of implementing biosecurity procedures swiftly.

Biosecurity protects both your flock and your family. Some illnesses can be passed from chickens to humans and other illnesses can be spread from your flock to neighboring flocks or avian wildlife. Therefore following proper biosecurity is of the utmost importance.

Implement biosecurity measures to prevent the spread of diseases and maintain a healthy flock include:

- quarantining new chickens before introducing them to the existing flock

- following sanitization protocols

- separating food and water

- not sharing equipment

- wearing disposable gloves and plastic shoe coverings

- washing your hands

- minimizing exposure (no visitors)

- calling a veterinarian

- alerting infectious disease control

- contacting neighboring homesteads, farms, or flocks

- maintaining clean and hygienic coop conditions (Jacob, n.d.-a).

Proper Sanitation and Hygiene

We have discussed the importance of regular coop cleaning and disinfection in various sections of this book. For more information on proper sanitation practices, you can refer back to Chapters 1 and 4.

Stress Management

Your flock and individual chickens may also experience times of stress. So, it is important to understand the impact of stress on chicken health and implement strategies to minimize stressors. Some strategies include providing a safe and secure coop environment, minimizing disruptions and loud noises, and ensuring appropriate space and resources for each chicken.

Dealing With Chicken Diseases and Pests

When dealing with viral or bacterial diseases, seek medical guidance before attempting any home remedies. Always follow veterinary guidance to make sure you get an accurate diagnosis and the appropriate treatment for your sick bird and to get advice on proper containment for possible outbreaks. Educating yourself on the most common illnesses in chickens and their symptoms can help you readily identify when there is a problem.

Recognizing Symptoms of Common Diseases

Knowing the signs and symptoms of these diseases may help with early detection and treatment, and prevent the spread of highly infectious contagions to your own and nearby flocks.

Below is an explanation of symptoms, treatments, and preventions for the five most common diseases (MannaPro, n.d.).

Fowl Pox

Fowl Pox presents itself in two different forms: wet and dry. In the wet form, you may see lesions around the chicken's mouth and discharge coming from its eyes. In its dry form, you may see wart-like lesions on bald patches or areas without feathers on your bird, including around the chicken's eyes.

Unfortunately, there is no treatment for this illness but it will likely resolve on its own. Quarantine your chicken and give them some extra care and attention while they recover for a few weeks. Prevention includes vaccinations, quarantine, and making sure your coop is nowhere near an area where mosquitos proliferate since these irritating insects help spread the wet form of fowl pox. Make sure to empty and remove any stagnating water and fill in any deep puddles to prevent mosquito breeding.

Infectious Bronchitis

Infectious bronchitis is a respiratory infection that can cause a decline in eating, drinking, and egg production. Additionally, you may notice your chicken has labored breathing, seems lethargic, and may or may not have discharge coming from its nostrils and eyes. There is no direct treatment for bronchitis except for time. Some preventative measures can be taken, like vaccinations and quarantining; however, like most avian illnesses, there is not much you can do.

Marek's Disease

Marek's Disease is a form of avian cancer that spreads through viral infection and typically affects young chickens around 12 to 25 weeks of age. Your chick is likely to have Marek's if they have partial paralysis, oddly shaped pupils, blindness, or tumors. Unfortunately, there is no cure or treatment for Marek's. If a chick survives the disease, it will be a carrier for its entire life causing death and illness to other flock members. The only preventative measure currently is to vaccinate newly hatched chicks.

Newcastle Disease

Newcastle Disease (ND) is a respiratory illness that causes difficulty breathing, nasal discharge, murky eyes, neck twisting, partial paralysis, and reduced egg laying. The different strains of ND vary in severity and mortality. Most adult chickens can recover without being carriers afterward but if a chick contracts the disease, they are likely to succumb to the illness and perish. There are no direct treatments except quarantine and extra care while the only preventative measure for ND is vaccination.

Be extremely cautious when caring for a bird with ND as the disease can spread to your entire flock via your clothing or even the shoes you were wearing while caring for the sick chicken.

Coccidiosis

Coccidiosis is a parasitic infection that chickens can contract from others through droppings and unhygienic coop environments. Symptoms of this parasitic infection include loose stools, watery or bloody diarrhea, lethargy, ruffled feathers, and weight loss. Contact your veterinarian for antiparasitic medication and possible antibiotics to prevent further infection. To prevent Coccidiosis, you need to keep your flock's environment very clean, make sure that food and water sources are uncontaminated by fecal matter, and add a good probiotic supplement to your chickens' daily food (Freedom Ranger Hatchery, n.d.).

Always seek veterinary guidance for accurate diagnosis and treatment.

Treatment Options and Medications

Understanding the different treatment options and medications available for chicken diseases can help reduce stress or anxiety and help you make informed decisions in the event any of your hens fall ill.

Medicating hens is not that different from medicating humans or house pets. Make sure you know how to safely use these medications before administering any of them to your chickens. Antibiotics are used to treat the symptoms of a bacterial infection in chickens and are occasionally used to prevent bacterial illness during times of immunocompromisation. Antiparasitics are used to eradicate parasitic infections, while antifungals are used to treat fungal infections.

Administering Medication Safely and Effectively

Medications can often be administered through your chickens' water or food; however, if they are very ill, they might not be able to eat or drink properly. If this is the case, you will have to administer the medicine to your birds individually via a syringe.

To use a syringe safely follow these steps (Lisa, n.d.):

1. Plunge the syringe into the liquid medication.

2. Slowly pull back the plunger of the syringe and carefully measure out the dose prescribed by your veterinarian.

3. Gently secure and restrain your chicken.

4. Place the syringe at the side of its beak.

5. Encourage your chicken to bite the syringe tip with the side of its beak.

6. Slowly administer the medication one small drop at a time.

Do not administer medication too fast or from the front of the beak, since this can cause your bird to choke.

Natural Remedies and Supportive Care

There are some fantastic natural and immune-boosting home remedies for treating sick chickens. However, make sure to administer these remedies under the supervision of an experienced veterinarian and never use them in place of any medication or treatment your vet has suggested.

Herbal Treatments and Supplements

Natural solutions play a vital role in poultry care since herbal treatments and supplements offer gentle yet powerful remedies for your flock (The Pioneer Chicks, 2023). Chamomile, calendula, and nettle tonics boost immunity, while garlic and oregano act as natural antibiotics. Lavender reduces stress, and essential oils like eucalyptus aid respiratory health. Herbs such as marigolds and basil repel pests, and probiotics from dandelion and parsley enhance digestion. Aloe vera promotes skin healing. Embrace these natural options for a healthier, happier flock, fostering a sustainable connection between your chickens and their environment.

Quarantine and Isolation Protocols

We have already mentioned quarantining in previous chapters, especially the situations in which you should quarantine your birds. However, there is more to quarantining than merely locking up your birds, if any of your birds get sick, you will need to follow strict protocols. So, let's look at the process step-by-step.

Appropriate quarantine measures for a sick chicken are as follows (Jacob, n.d.):

1. At the first sign of illness immediately separate your chicken from the flock.

2. Set up a small dimly lit enclosure such as an appropriately sized pet carrier.

3. Do not allow the enclosure or your sick chicken to be anywhere near your flock.

4. Ensure the enclosure is kept in a quiet room away from any noise so that your bird can rest comfortably without the risk of getting startled.

5. Place some comfortable bedding inside the enclosure.

6. Provide your chicken with easily accessible food and water.

7. Place your chicken in their quarantine enclosure.

8. Call your veterinarian and set up a time to get your chicken checked out.

9. Deep clean your coop, remove any old bedding, and flooring, and clean all food and water sources thoroughly.

10. Monitor your quarantined hen every few hours and be sure to keep an eye on the rest of your flock to ensure the illness has not spread.

11. Do not re-introduce your hen to the rest of the flock until your vet gives you the okay to do so.

Other protocols to keep in mind are:

*   Wash your hands immediately after handling a sick chicken.

*   Do not touch your face before washing your hands.

*   Do not handle other members of your flock after caring for your sick hen.

*   Do not allow small children or anyone whose immune system is compromised near any sick chickens.

- Do not consume the eggs or meat of a sick chicken.

Following these procedures can prevent the spread of illness and harm to both your flock and the rest of your farm.

Supportive Care for Sick or Injured Chickens

The best way to support a sick or injured hen is to make sure they are calm, secure, well-hydrated, and have food readily available.

If your chicken is injured, it may be necessary to swaddle them to avoid further injury. Make sure that your chicken remains calm and does not show any signs of distress.

Interactive Exercise

Develop a flock health checklist and perform a comprehensive health check on your chickens. Monitor their overall well-being, note any abnormalities or concerns, and keep a record of their health history. Implement necessary preventive measures and interventions based on your observations and research.

Chapter 7:

Chicken Behavior and Communication

Any chicken keeper will tell you how important effective communication between you and your flock is in helping it thrive. This chapter will help you understand chicken behavior and communication and empower you to interpret their needs, emotions, and social dynamics, leading to healthier flocks. Effective communication fosters trust and cooperation, ensuring their well-being. Whether you are a seasoned poultry keeper or a novice, we will equip you with tools to confidently navigate chicken behavior, forging a deeper, mutually beneficial connection with your birds. Explore the hidden facets of your feathered companions for healthier eggs, affection, and harmonious coexistence.

Understanding Chicken Behavior

You will not necessarily form an instant connection with your chickens; it is more like a well-crafted friendship that develops over time. These feathered beings have their own personalities and quirks and need time to get accustomed to their new environment and your presence and learn to trust you. It is a gradual process that involves understanding their behavior, preferences, and idiosyncrasies.

Just like with any relationship, patience is key. So, take the time to observe their social interactions, their habits, and how they respond to your actions. Over time, you will learn the subtleties of their body language, the nuances of their vocalizations, and what makes each chicken unique.

As you invest this time and effort, you will become attuned to their needs and emotions and your chickens will start recognizing you as a friend, a provider, and a guardian. This bonding process is not just about you getting to know them but also about them getting to trust and rely on you. So, do not rush it; enjoy the journey of discovery and connection with your feathered companions. It is a rewarding path that leads to a stronger and more fulfilling relationship.

The Social Structure and Hierarchy of Chicken Flocks

Chickens are naturally social creatures that thrive on order and interaction. While we have grazed on the topic of pecking orders in various sections of this book, we have not fully delved into the intricacies of chicken politics. Pecking orders are key to understanding the instincts and behaviors of the chickens in your life.

Pecking orders naturally start to develop when chicks are merely a week old. By the time they reach six weeks of age, a clear and indisputable pecking order will have been established. Fascinatingly, this process is sorted out through play. When chicks play together they perform 'playful fights' through friendly sparring, and this process is what helps them establish their pecking order.

Adult flocks that have hatched together typically live in harmony within their hierarchical social system. Chickens are capable of recognizing the members within their flocks and know exactly where in the order they stand from the strongest, biggest, or most dominant chicken at the top to the docile pushovers at the bottom.

Not only do chickens have special rankings but they also have specific behaviors based on rank. Lower-ranking birds show signs of submission to higher-ranking members by running away, crouching, or even squatting. Oddly enough, chickens may consider their keeper a part of their flock and may even demonstrate signs of submission towards them—this may be especially true if you hatched your chicks (McCrea & Baker, 2022).

As previously discussed, when you suddenly add a new chicken to the flock, it disrupts the pecking order and can lead to chaos. The flock will either form a new social structure over time or will, sadly, end up mauling the new bird—if you want to know how to manage the social integration of a new bird refer back to this section in Chapter 2. However, this is not the only dilemma that comes with introducing a brand new chicken to your established flock.

Chickens are social learners, meaning they learn new behaviors by watching other flock members, pets, and even humans perform the behavior. If your new adult chicken was raised elsewhere it may come with some undesirable habits or behaviors. This can spell trouble if the new hen happens to fall on the dominant end of the pecking order. Once they establish their dominance, the chickens under them will start following and integrating the newcomers' behavior—for better or worse.

However, social learning does not stop there, it also forms part of a chicken's survival skills. Chicks will learn from their mothers what to eat and where the good food is, as well as where their territory boundaries are and where they can find a safe place to perch (McCrea & Baker, 2022). So, if your flock newbie is a picky eater and reacts with disgust to healthy veggie treats or your selected feed, you might be in for a headache.

Recognizing Natural Behaviors and Instincts

Despite appearances, chickens are marvelously complex. Like all creatures, they have innate instincts that have been developed over countless years of evolution. Instinctive behaviors include preening, 'fight or flight' responses to perceived danger, scratching the ground, and of course, listening to their mother. Oddly enough, behaviors such as drinking or knowing how to eat food, are not instinctual. Chicks must be taught these things by the mother hen or by the keeper who hatched them.

Chickens have a variety of natural behaviors that can be separated into a few categories, namely, maintenance behaviors, exploration behaviors, and social behaviors.

Maintenance behaviors are life essentials that sustain the chicken's physiological equilibrium. A large portion of this type of behavior revolves around foraging and feeding, which chickens spend more than half of their time awake doing. Chickens have an innate need to work for their food—even when resources are easily available. Some natural behaviors of foraging include scratching and pecking at the ground for tasty bugs, grass, edible flowers, and tiny pebbles.

Feeding and foraging is also a social activity, so you may notice that once one chicken starts feeding the rest of your flock will follow. If a chicken is not able to forage or if enough run space is not available, your chickens may exhibit some negative behavioral patterns including, aggressive preening or feather plucking, egg eating, and even cannibalism (McCrea & Baker, 2022).

Chickens need rest and sleep just like humans do, and they especially enjoy roosting on any perch-like surfaces.

Interpreting Chicken Vocalizations and Body Language

Chickens are quite the communicators, and they have two main ways of getting their point across: body language and their voices. Think of it like this; when a chicken wants to chat with their flock up close, they strike a pose or give a little shimmy. But when they want to shout out to the whole neighborhood, they unleash their vocal talents, and their voices can carry for quite a distance, even reaching a neighboring flock. So whether they are chatting to their coop-mates or belting out the latest chicken gossip to the entire barnyard, these feathered friends know how to keep the conversation going.

Chickens have quite the repertoire when it comes to expressing themselves. They are like little feathered actors, using their bodies to tell stories. They may tilt their heads at funny angles, raise or lower their tails, and even fluff up or flatten their feathers to convey messages about personal space, health, and how they are feeling in the group. It is like a chicken soap opera, and these movements are their way of making sure everyone knows exactly what is going on, especially during dramatic moments like mating rituals and territory tiffs.

Chickens have their own language, and it is like music to their ears—and ours, if we are listening closely. Just like we chat with words, chickens have their own unique vocalizations and language. When a rooster lets out that classic crow, it is not just a wake-up call; it is his way of saying, "This is my turf!" like a feathery version of "Stay off my lawn" without the need for feathers to fly.

The chicken vocabulary is far from limited. With over thirty vocalizations in their repertoire, each has its own unique meaning. From expressing contentment to shouting "danger!" or even announcing their romantic intentions, they have a sound for it. Chickens may give a little shout to alert others of food and express excitement, fear, or even surprise. When one chicken chirps, the others understand perfectly and even recognize members of their flock based on their voices. For instance, if a little chick lets out a distress call, mama hen goes into superhero mode and rushes to the rescue.

The importance of vocalization is made clear by the fact that chickens start communicating even before they hatch. Pre-hatched chicks are little chatterboxes inside their warm and cozy eggs and they like having conversations with their mom and even with their fellow egg mates. It is almost like they are synchronizing their watches to make sure everyone pops out of their eggs at the same time.

Building a Bond With Chickens

Developing a strong bond with your chickens will help you manage the flock and take care of their well-being. Bonding helps you understand your birds and identify any health issues promptly, it also reduces stress, enhances health, and streamlines care routines. Beyond these practical benefits, it brings personal satisfaction and a deeper connection with your flock, making it a mutually rewarding endeavor.

A strong bond with your chickens fosters a sense of companionship. Chickens can recognize their caregivers and become more responsive to their presence. They are more likely to trust you, making tasks like handling and medical check-ups easier. This bond also creates a harmonious environment, as happy, contented chickens are more likely to be productive layers and cooperative members of the flock. In essence, bonding with your chickens goes beyond mere animal husbandry; it creates a fulfilling, enriching experience for both you and your feathered friends.

Building Trust and Fostering Positive Interactions

There are many ways to build a bond with your chickens. A few of these include:

- spending time with your chickens

- teaching them to recognize your voice

- bringing tasty snacks and treats

- respecting their boundaries

Building trust with your flock takes time and patience. When dealing with chicks, it is essential to establish a connection from day one. Spend quality time with them, gently cradle them, and have one-on-one sessions daily. Allow them to become familiar with your presence, chat with them softly, and you might even find them dozing off in your hands or on your lap.

For your more mature hens, bring along a comfy seat and set it up near their stomping grounds. Get down low to their level so you do not give off any intimidating vibes. Try to keep your noise levels low and move without any sudden jerks. Curiosity will eventually get the best of them, and they will mosey over to investigate your presence. If they seem a bit shy or skittish, you can sprinkle some feed around to sweeten the deal and coax them closer.

On your visits to care for your feathered buddies, it is good practice to announce your arrival. Gradually, they will come to recognize your voice and link it with the prospect of a tasty meal and some companionship. Do not shy away from indulging in some light chit-chat with them while attending to your tasks. Remember, a calm and unhurried demeanor is the key to forging trust. As you replenish their feed, take a leisurely stance nearby and observe their behavior.

Chickens adore treats, whether it means getting leftovers from your kitchen or a juicy piece of fruit. You can easily become their treat-time hero while building trust and rapport. Start by ensuring you are present when you offer treats, making their mealtime an enjoyable experience with your company. Gradually introduce a variety of treats to expand their palate, and if they eagerly nibble, it is a sign that trust is growing. Once you have established trust, entice them to your lap with the promise of more treats, but be mindful of their curious beaks! While they enjoy their snacks, gently pet them to enhance your bond with each chicken. This journey of treats is a delightful path to forging a closer connection with your feathered friends.

Building trust with your chickens involves understanding their individual preferences and boundaries. Chickens, like humans, have different personalities; some may enjoy being petted, while others might prefer to maintain a bit of distance. Additionally, their temperament can be influenced by their breed. For example, breeds like Barred Rocks are often known for being more sociable and open to human interaction (Ranson, 2021).

To build trust effectively, observe your chickens and learn which ones are comfortable with physical contact and which ones need more space. By respecting their individuality and boundaries, you can establish a strong bond with your feathered friends, creating a harmonious environment in your coop.

Training Chickens for Basic Commands or Behaviors

While the idea of chicken training might raise a few eyebrows, in reality, chickens are very clever and it is possible to teach them simple tricks—like walking through obstacle courses, following commands like jumping up, laying down, and even wearing a walking harness. Many keepers succeed in training their chickens using reinforcement techniques, which can either be positive or negative depending on the situation. Learning to train your chicken is not only fun, but it is also a handy tool for adjusting any unwanted behavior your feathered friends might be exhibiting (Keyes, 2020).

Engaging Chickens in Enriching Activities

We have given some brief examples of enrichment activities in various locations in this book, but not in any detail. By adding enrichment you can help keep your chickens happy and a happy chicken is a friendly chicken.

Enrichment activities can help you bond with your chickens by reducing negative emotional states, boredom, bullying, and feather pecking. Enrichment provides essential cognitive stimulation and provides your flock the opportunity to engage in natural behaviors.

There are two types of enrichment you can provide to your flock, social enrichment and environmental enrichment.

Social enrichment simply means providing your chickens with some company. Chickens are flock animals and therefore need the company of other feathered friends. Keeping a single chicken all by itself will ultimately lead to distress and self-harm. So bear this in mind when planning your flock size.

Environmental enrichment comes in the form of toys and activities. Just like cats and dogs need toys and fun exercises to keep them occupied, healthy chickens require the same kind of entertainment.

Below is a list of activities and toys you can add to your flock's environment (BCSPCA, 2021).

- Jungle gyms and elements such as, swings, teeter totters, ladders, ramps, and platforms

- Dust baths are not only important for physical health by preventing external parasites but are fantastic social and environmental enrichment activities.

- Odds and ends, including objects like pinwheels, bells, and bird-safe mirrors—which can be obtained in any pet store with a parrot toy section.

- Dig boxes are like a sandbox full of treats hidden inside. You can make a dig box by enclosing dirt in an open space, like an old kiddie pool, sandbox, or old tire. Hide treats in the dirt for your chickens to scratch up and enjoy!

- Treat balls are round hollow balls stuffed with treats for your chickens to roll around, as they peck and play with the ball treats will fall out.

- Puzzle feeders are fun for animals of all shapes and sizes. You may have seen these in the pet toy aisle or watched a dog try to get their treat out of one, but did you know that chickens love solving puzzles too?

- Vegetable swings are easy enrichment toys to make. Whether you bundle up an assortment of veggies with some kitchen chords or simply tie a cabbage with some rope and hang it in the coop, your chickens are bound to get a kick out of it!

Dealing With Behavioral Challenges

We have discussed the various quirky and fun personality traits that chickens exhibit as well as their political hierarchy of pecking orders and while these qualities are quaint, there is a dark side to chicken social dynamics, including aggression, waking the neighbors, broodiness, and even cannibalism.

Addressing Aggression, Bullying, or Pecking Order Issues

Aggression and feather issues take the spotlight as the most common behavioral challenges in a flock. These two troublemakers share common roots, like stress, cramped living spaces, and competition for essentials like food and water. The good news is that these problems can often be kept in check by tackling the underlying causes and, in some cases, giving the boot to the main troublemakers. Mixing things up by introducing some enriching activities and reshuffling the social pecking order can do wonders as well. Now and then, things can take a dark turn towards cannibalism, especially among broilers and free-range flocks, where these problems tend to crop up more often due to living conditions. Egg-laying hens in large-scale production do not usually encounter these issues since they live in smaller, more controlled groups.

Aggression often wears the mask of head and face pecking or feather plucking, and with those sharp beaks, it can lead to some pretty nasty injuries. To calm the waters, you can make use of tricks like reducing the amount of daylight in battery setups, adding a dash of tryptophan to their meals, and, as a last resort, trimming the beaks of the offenders. However, beak trimming is not a long-term fix and can raise some welfare concerns, so rather try the other options first before considering it.

On a brighter note, chickens are all about personal grooming and feather care. They take pride in it, you could say. Sometimes, it is even a social event! And if you ever spot them dustbathing, do not worry; they are not practicing for a chicken Olympics. It is just their way of keeping those feathers neat and tidy and, believe it or not, reducing the chances of pecking squabbles (Landsberg & Denenberg, 2022).

Feather pecking and cannibalism cast a dark shadow over poultry keeping. Cannibalism often emerges from aggressive behavior among chickens and other fowl and sometimes starts with the dominant members of the flock pecking the feathers of their less dominant flock mates. However, it can escalate into more sinister acts, including pecking the vent of a hen immediately after she has laid an egg or even targeting the skin on the head, comb, wattles, or toes. Cannibalism is the grimmest outcome of these pecking behaviors, resulting in severe harm to the victimized birds.

The roots of cannibalism are complex and it is difficult to pinpoint a single cause. Rather, it is a web of factors that contribute, including genetics, overcrowding, intense lighting, and dietary imbalances. It is worth noting that overweight pullets entering egg production or hens in full production face unique risks, as the mucosa protruding from the vent during and after egg laying can become a fatal attraction for pecking. A shortage of feeder space, deficiencies in essential minerals and vitamins, skin injuries, and the failure to promptly remove deceased birds from the flock all add fuel to this destructive fire.

Beyond the immediate harm, cannibalism often acts as a conduit for infectious diseases like erysipelas and botulism, further jeopardizing the flock's well-being (Crespo, 2023).

Clinical signs of cannibalism are most prevalent in floor systems where eggs are laid on the ground in crowded spaces. Birds victimized by this cruel behavior typically exhibit damaged and ragged feathers, often accompanied by a drop in egg production. Furthermore, flocks experiencing vent pecking may see a surge in prolapse cases, and overall mortality rates can soar due to the relentless attacks.

Upon examination, affected birds may bear the scars of their suffering, including poorly feathered bodies and torn, injured flesh. In severe cases, the damage can lead to hemorrhaging, compounding their misery. The presence of blood on exposed skin can serve as a gruesome invitation for more pecking, ultimately resulting in the bird's demise. Vent pecking typically occurs right after the hen has laid her egg since it is the exposed mucous membrane that triggers the aggression of fellow birds.

Controlling and preventing this harrowing issue hinges on addressing the underlying risk factors. As soon as you detect any signs of pecking or cannibalism, you need to intervene swiftly, as once this destructive behavior becomes entrenched, it is exceedingly challenging to eradicate. Strategies include rectifying dietary deficiencies, substituting mash feed with pelleted feed, opting for floor litter rearing rather than slats, reducing the intensity of lighting, and providing perches to offer respite for targeted birds.

You can also provide more environmental enrichment, such as hanging white or yellow strings, which may prove beneficial. Additionally, you can consider beak trimming, which involves carefully trimming the sharp tip of the upper beak to minimize skin trauma caused by pecking. This procedure can be performed as early as day one of the bird's life and repeated between 6 and 12 weeks in maturing pullets or turkeys. For older poultry, cauterization is necessary to ensure proper hemostasis during beak trimming, all in the name of safeguarding their well-being (Crespo, 2023).

Managing Excessive Noise or Disturbance

Chickens are charming chatterboxes and despite their petite size, they possess a remarkable talent for making themselves heard. Unfortunately, our dear clucking companions can ruffle nerves as well as feathers, especially when they decide to hit a few high notes at night.

An unwanted midnight concert can lead to sleep disruptions and even disgruntled neighbors. But humans are not the only ones who get disturbed by loud chickens. Excessive coop nose can interfere with your flock's social lives. Much like an opera gone awry, squawking at all hours affects chicken conversations and results in more noise, upheaval, and a decline in egg production. Drama, everywhere!

Fortunately, if your chicken chatter is getting out of hand and disturbing the peace—for all involved— there are a few things you can do to reign in the clucks and cackles.

To make sure that you maintain a calm environment for your flock, you first need to provide an appropriate shelter. Without refuge from the unpredictability of the weather and ample room for stretching and flapping, chickens can get incredibly rowdy. A comfortable, draft-free coop works wonders in keeping the flock calm and stress levels down, which is a huge factor in reducing noise. Having too many chickens in a small space is a surefire recipe for drama and amplified vocal expressions. So, if your flock is expanding but their coop is not, it might be time to reevaluate their living conditions.

Another way to get rowdy chickens to quiet down is to make sure they are getting the nutrients they need including vitamins, minerals, carbohydrates, proteins, and fats. If your chickens are getting the nutrition they need but are still determined to make a ruckus, add some herbal zen to their feed or snacks. Calming herbs like mint and oregano can have calming effects on your chickens and reduce their noise levels.

If none of this works, try providing resources for your hens to get their energy out like adding forage toys, swings, and even expanding your run space (Azeem, 2023).

Modifying Behaviors Through Positive Reinforcement

We already know that training your chicken is not only possible but can be a great way to mitigate poor behavior. So, let's discuss positive and negative reinforcement training a little more.

Positive reinforcement means giving your chicken a treat right after a behavior or action you want to encourage. Doing this is like saying "good job" with a tasty reward. This positive reinforcement makes your chicken more likely to repeat the behavior.

On the other hand, negative reinforcement means taking away something unpleasant when your chicken does the right thing. It is like creating a comfy, happy zone for them when they behave the way you want them to.

Studies across different species, including humans, dogs, and yes, chickens, show that positive reinforcement works better than punishment for shaping behavior (Keyes, 2020). So, in your chicken training adventures, keep those tasty treats handy and watch those good behaviors flourish.

Tips for Dealing With Broody Hens

Every hen has the potential to be broody since broodiness is a natural part of chicken life. This overwhelming instinct is what leads hens to want to nest and hatch a few chicks—which are also referred to as clutch. We do not know precisely what triggers broodiness but it is likely a combination of hormones, instinct, and maturity—not unlike what triggers the baby blues in young couples (British Hen Welfare Trust, 2023).

Understanding Broodiness

Just because a hen has become broody does not mean that she is going to hatch any chicks. In fact, some hens are just naturally more maternal than others and may get stuck in a brooding cycle without laying eggs (British Hen Welfare Trust, 2023). This can have negative health impacts since the hen might not be eating or drinking enough, which can lead to dehydration and malnourishment, and your hen may stop laying eggs altogether.

Recognizing Signs of Broodiness

Spotting a broody hen is relatively straightforward. You might find your hen staking out their favorite nest box and getting flustered and worked up if you try to disrupt it. Expect ruffled feathers and, on occasion, even a few angry pecks if you attempt to intervene or remove her from her nesting box. If you happen to have a rooster, some other obvious signs of brooding might include mating rituals.

The Impact of Broodiness on Egg-Laying

If a hen is laying a clutch, it is not uncommon for her to lose feathers and even develop a bald patch on her belly. Do not worry, this is completely normal. The bald spot helps to provide extra warmth to help incubate her precious chicks. However, if your hen is not laying fertile eggs then the balding is for naught.

Since your hen is prioritizing her eggs, she might need some help keeping up her strength. Lift your brooding hen out of her nest each day so she can eat, drink, and keep her strength up. Even if she immediately returns to the nest after drinking and eating, this brief break provides her with a bit of exercise and a little is better than nothing.

How long does brooding last? If she is truly hatching eggs, the hen will stay in the nest for about two weeks, but her broodiness might linger for up to six weeks. To help break this cycle, consistently cool her underside by periodically removing her from the nest.

If you would rather let nature take its course, remember that eggs require around two to three weeks to hatch. During this time, ensure that the hen has access to food and water, which you might need to encourage her to find by periodically removing her from the nest box, as mentioned earlier. If there are other hens in the nesting box, she should be willing to share, but keep an eye out for potential aggression (British Hen Welfare Trust, 2023).

Strategies to Break Broodiness

The easiest way to prevent and fix a brooding hen is by removing her from the nesting box and collecting her eggs regularly, even if she resists moving. If she keeps returning to the nest box, consider blocking it off by placing a piece of wood in the entrance. Another approach is to put ice cubes in her nest box. This makes it uncomfortable for her to sit on and reduces her body temperature. When hens get broody, their body temperature tends to rise, so this cooling effect can help reduce their broodiness.

If you have a particularly territorial hen, you might have to confine her outside the coop temporarily, to collect her eggs.

However, collecting eggs might not stop this behavior. In chronic cases, some keepers find fertilized eggs for their broody hens. After all, she just wants to be a good mama. Unfortunately, there is no way to tell if the fertilized eggs are hens or roosters. So be advised that you might end up with quite a few male chicks, which might not align with your intentions.

Should your hen remain persistently broody, remove her from the coop entirely and place her in a cage with a flat bottom. Remember to provide food and water but no bedding. In this way, you are creating a space that is intentionally uncomfortable and should help reduce her broodiness. This usually only takes a few days, but the timeframe can vary from case to case. You will know when your chicken's broodiness has subsided when she no longer fluffs her feathers and shows no eagerness to return to the nest (British Hen Welfare Trust, 2023).

Interactive Exercise

Conduct a behavior assessment on your chickens. Identify the behavior traits or personalities of each chicken, and identify issues such as aggression, broodiness, or issues with fitting into the flock. Develop a plan to address these challenges and implement measures to minimize the risk of harmful behavioral problems escalating.

Chapter 8:

Maximizing Egg Production

Eggs are not just a staple on your breakfast table; they are also a valuable resource that your feathered companions provide you with. In the pages that follow, we will delve into a world of strategies and techniques meticulously designed to coax your hens into embracing their peak egg-laying potential. From understanding how a chicken's breed and age impact egg production to ensuring that their nutritional needs are met, this chapter will guide you through every step you need to ensure a consistent and bountiful supply of fresh, farm-to-table eggs.

Understanding Factors That Affect Egg Production

The main factors that affect egg production are:

- age

- seasonal changes

- brooding

- lighting

- stress

- breed

- health (Purina Mills, n.d.-a)

Age and Breed Considerations

Understanding how age and breed influence egg production, can help you determine which breeds to add to your flock and when it is time to add new birds. By making informed decisions, you are well on your way to an egg-citing and rewarding poultry-keeping adventure. Whether you are aiming for abundant egg yields, a balance between eggs and meat, or simply the pleasure of delightful, feathered companions, these insights will guide you on how to have a consistent supply of farm-fresh eggs and an enriched chicken-keeping experience. So, get ready to appreciate the diverse characters and culinary delights your hens bring to your life!

The Peak Egg-Laying Period for Different Breeds

The breed of chicken you choose will determine the level of egg production that you can expect. Different breeds have distinct characteristics, including their tendencies for laying. Some breeds are celebrated for their prolific egg-laying capabilities and are ideal choices for those seeking a high egg yield.

The peak egg-laying period can also vary significantly depending on the breed you choose. Poultry keepers looking to optimize egg production therefore need to have a good understanding of breed-specific patterns and capabilities.

For high-egg breeds like the White Leghorn, Rhode Island Red, and Sussex, the peak egg-laying period typically falls between six to eight months of age (FIA, n.d.). During this time, these hens are like egg-laying machines, consistently churning out large quantities of eggs. It is not uncommon to collect an egg from each of these hens almost daily during their prime.

On the other hand, dual-purpose breeds such as Plymouth Rock and Orpington exhibit a slightly different pattern. While they may start laying eggs at around six to seven months, their peak period is known to last longer (Rachael, 2023b). If egg-laying breeds are sprinters, then dual-purpose breeds are marathon runners and often maintain a steady egg-laying rate well into their seniority. So, while the initial egg production might not be as intense as high-egg breeds, they compensate with more consistent productivity throughout their lives.

Specialty breeds like Silkies or Polish chickens, known more for their captivating looks and charming personalities than egg production, have their own pattern. They usually begin laying eggs at around seven to eight months but their laying habits tend to be more sporadic and unpredictable (Rachael, 2023b). While you might not rely on them for a daily egg, their occasional surprises can add some excitement to your basket.

Recognizing Changes in Egg Production as Hens Age

Age is a pivotal factor in egg production, and the various stages of a hen's life influence her laying capacity. Pullets are young hens typically less than one year old, and they represent a phase of anticipation and eagerness. While some pullets may start laying eggs at as early as five months, their production tends to be sporadic and the eggs they produce are generally smaller (Purina Mills, n.d.-a). However, as pullets mature and their reproductive systems fully develop, they begin laying on a more consistent basis.

Around the age of six to eight months, hens usually enter their prime egg-laying period (Purina Mills, n.d.-a). This phase can extend for several years and is characterized by remarkable consistency in egg production, often resulting in an egg each day. During this peak phase, you can expect the highest quantity of fresh eggs from your flock, and these eggs are typically of ideal size and quality.

As hens age beyond three years, their egg production often starts to decline slightly (Purina Mills, n.d.-a). While some older hens may continue to lay, the frequency and quality of their eggs may decrease depending on the breed. However, senior hens offer other valuable contributions to the flock, such as their experience and maturity, which make them exceptional caregivers and able to provide stability to the group. Many chicken keepers find value in retaining senior hens for these reasons, even as they anticipate a reduction in egg production.

Seasonal Variations

All poultry keepers need to understand how seasonal changes affect egg production. Chickens, like many animals, respond to the changing seasons, and understanding these responses helps you prepare for changes and maximize production.

As spring arrives, longer daylight hours and rising temperatures trigger an upswing in egg production. This season is marked by heightened reproductive activity and abundant egg-laying. Summer, characterized by ample daylight and favorable weather, sustains consistent egg production, with chickens staying active and well-fed, contributing to a steady flow of eggs.

However, with the onset of fall and shorter daylight hours, egg production will gradually start to diminish. Chickens are sensitive to decreasing sunlight, leading to a decline in their reproductive activity. Nonetheless, many hens continue to lay eggs throughout the fall.

Winter poses the greatest challenge to egg production, as reduced natural light can cause a significant drop in egg laying, with some hens ceasing production entirely. To maintain a steady egg supply, many chicken keepers add artificial lighting to their coops to supplement the diminishing daylight.

In summary, recognizing how seasons impact egg production enables poultry keepers to adapt their care routines, ensuring a consistent egg supply throughout the year. Chickens' egg-laying patterns are strongly influenced by the shifting seasons, and this understanding helps maintain optimal egg production.

The Effect of Daylight Hours on Egg-Laying

Chickens operate on a distinct biological rhythm, much like humans. To maintain peak egg production, they require approximately 12 to 14 hours of daylight. This demand is comfortably met during the summer months when days are longer. However, as winter approaches and daylight diminishes, your hens may experience a significant reduction in their laying activity. Instead of the usual four eggs per week per bird, you may find yourself collecting between one and three per week (Spencer, 2021).

Strategies to Maintain Consistent Egg Production Throughout The Year

The drop in egg production during the winter months can be a concern for poultry farmers and egg enthusiasts. To address this issue and maintain consistent egg production throughout the year, they often turn to artificial lighting solutions within the coop. By doing so, they provide their chickens with the illusion of extended daylight hours, even during the darkest days of winter.

Here are some ways you can effectively use artificial lighting to manage egg production:

- Install timers: Make sure to regulate the artificial lighting in your coop using a timer to restrict exposure to a maximum of 12 hours of light daily. This balance allows your chickens to continue producing eggs without compromising their health or sleeping patterns.

- Safe lighting placement: Position the light source safely out of the chickens' reach to prevent pecking, potential damage, or worst-case scenarios, such as fires.

- Natural light: If you prefer a more natural approach, consider incorporating robust, insulated windows into your coop's design. These windows will enable your chickens to benefit from available natural light, ultimately supporting their physical well-being and overall comfort during the colder months.

By understanding how lighting affects egg production and implementing these strategies, you can maintain consistent egg production all year long while ensuring your flock thrives and stays healthy.

Health and Well-Being

In Chapters 4, 5, and 6 we discussed the importance of maintaining the health and wellness of your flock and how certain illnesses can affect your chickens' egg-laying ability. To learn more about keeping your chickens in tip-top shape, you can refer back to those chapters. Below is a brief overview and reiteration of some of the subjects covered previously as they are extremely important to your chicken-keeping journey.

Proper Nutrition, Parasite Control, and Stress Management

Maximizing egg production hinges on several key factors, including proper nutrition, effective parasite control, and adept stress management. If you want to maintain a flourishing egg production operation you will need to pay meticulous attention to your flock's dietary needs. Providing a well-balanced diet rich in calcium, protein, and essential vitamins is akin to fueling your chickens' egg-laying engines. Insufficient nutrition can lead to eggshell fragility and a decline in egg output.

Additionally, you will need to exercise vigilant parasite control to sustain egg production. Unwanted guests like mites, lice, and intestinal worms can compromise your chickens' health and their ability to lay eggs. However, through regular inspections and by promptly applying suitable treatments you can curb these infestations.

Stress management also plays a critical role in optimizing egg production. Chickens, like all living beings, thrive in a stress-free environment. Overcrowding, extreme weather conditions, potential predators, or abrupt changes in their living environment all induce stress and disrupt their egg-laying patterns. To counteract this, provide ample living space, secure shelter, and a stable environment that helps to maintain lower stress levels.

A steady supply of fresh water is paramount for chicken health in general but it also ensures a smoother egg-laying process. So do regular health check-ups and vaccinations, since these are indispensable in preventing diseases and ensuring healthy egg production.

In summary, to establish a thriving poultry operation, you will need to understand how to maximize egg production through a range of methods, the importance of proper nutrition, as well as how to employ parasite control and cultivate a stress-free environment.

Ways to Promote a Healthy Environment for Optimal Egg Production

As long as your chickens are happy, safe, and well-nourished you should not experience any issues with overall egg production. However, for any keeper who prefers to have lists of what to do, here is a complete one.

- Provide adequate living space for your chickens to reduce stress and prevent overcrowding.

- Ensure your coop is well-ventilated to maintain proper airflow and prevent the buildup of harmful gasses.

- Keep nesting areas clean, dry, and free from feces and debris.

- Regularly inspect and treat your chickens for parasites like mites, lice, and worms.

- Maintain a consistent and balanced diet for your flock, rich in calcium, protein, and essential vitamins.

- Supply fresh, clean water daily to keep your chickens hydrated and support egg production.

- Implement predator-proof measures to protect your flock from potential threats.

- Minimize sudden changes in the chickens' environment or routine to reduce stress levels.

- Provide a comfortable and secure shelter, especially during extreme weather conditions.

- Conduct regular health check-ups and vaccinations to prevent diseases that can affect egg production.

- Create a calm and peaceful environment with minimal disruptions for your chickens to thrive in and lay eggs consistently.

Encouraging Hens to Lay More Eggs

Providing the Ideal Nesting Environment

Creating a comfortable and attractive nesting space for your hens will encourage them to lay more regularly. In Chapter 2, we discussed the importance of nesting boxes, the best way to set them up, and what materials to fill them with. In this section, we will provide a brief overview of nesting boxes and provide insight into how they impact egg production.

Nesting boxes protect eggs from dirt, feces, and breakage, while bedding materials provide a cushion for the fragile eggshells and reduce the risk of damage. Other advantages of nesting boxes include:

- Routine: Nesting boxes establish a consistent location for egg-laying. Hens are creatures of habit and tend to return to the same nesting box to lay their eggs every day. This predictability makes egg collection more efficient and ensures that your hens are laying their eggs in a place where you can easily retrieve them.

- Minimizing stress: An organized and designated egg-laying area reduces stress among the flock. When hens need to compete for nest sites, they become stressed or anxious, which means they may lay fewer eggs or stop laying altogether. Nesting boxes help maintain a calm and harmonious environment for the whole flock.

- Regulating temperature: Nesting boxes can help regulate the temperature of the eggs. Eggs are sensitive to temperature fluctuations, and laying them in a well-insulated nesting box protects them from extreme heat or cold, ensuring a higher ratio of viable eggs.

- Easy Egg Collection: For chicken keepers, nesting boxes make egg collection more convenient. Eggs are concentrated in one area, making it simpler to retrieve them without disturbing the entire flock.

- Managing broodiness: If you have broody hens (those that want to sit on eggs and hatch chicks), nesting boxes can help keep them from taking up all the laying space. Allowing other hens to continue laying in their own boxes.

Nest Boxes: Size, Location, and Materials

When it comes to designing nesting boxes you will want to create a comfy, welcoming space for your hens. Selecting the right nesting box size will help keep your hens happy and ensure a harmonious coop. Happy hens are known to be quieter, and who would not want a peaceful coop in their small space?

Nesting boxes are designed to be clean and comfortable and are often lined with soft bedding materials like straw, hay, or wood shavings. They should also be easily accessible to each member of the flock to prevent your hens from sneakily making their nests in odd corners of the coop or from laying their eggs in difficult-to-find places in the case of free-ranging birds. For more information about bedding materials and boxes take a look again at Chapter 2.

The ideal size of your nesting boxes may vary depending on a variety of factors including breed, quantity of eggs laid by your flock, the size of your coop, and whether or not your brooding hen is incubating a clutch.

Make sure there is enough space in the nesting boxes for enough bedding to provide adequate insulation so that your hens can comfortably sit and lay their eggs.

For standard-sized breeds, such as Rhode Island Reds or Plymouth Rocks, a good rule of thumb is to provide nesting boxes that are approximately 12 inches wide, deep, and tall—essentially a square-foot box (Winger, 2023).

However, if you have larger breeds like Jersey Giants or Orpingtons, you might want to offer slightly larger nesting boxes to accommodate their hefty size, which means 14 inches in width, depth, and height should suffice.

Dainty Bantams on the other hand might be just fine in a 12-inch box since they prefer things a bit cozier. A nesting box that is 10 inches in width, depth, and height is typically enough to keep them content.

What about brooding hens? If you have a hen who has just laid a clutch, it is helpful to provide her with a nesting box to accommodate both the new chicks and her brooding behavior. For a standard breed, a 14-square-inch nesting box should be large enough to keep her comfortable and reduce stress during this difficult stage of her motherhood.

If your flock is a mix of different breeds (or species), you have the option to build an array of different nesting box sizes to accommodate each member of the flock or build community nesting boxes with adjustable dividers. This way, all your hens have space according to their individual needs, ensuring everyone has a comfy spot to lay their eggs.

Bedding Cleanliness

Make sure to replace bedding daily for both comfort and hygiene. If your chicken is brooding, try not to disturb her nest too much—there is precious cargo in there. Remove any soiled portions and add a fresh layer to her box. If her nest becomes wet, due to an unforeseen leak or accident, carefully and gently remove the fertile eggs and replace all the bedding with fresh, clean layers.

Offering Privacy and Security for Hens While They Lay

Hens usually prefer to lay their eggs in a secluded and secure spot, away from any possible threats. Nesting boxes provide a private area that mimics the safety of a hidden nest in the wild, where a hen can lay her eggs knowing that both she and her eggs will be safe. When hens feel safe and undisturbed, they are more likely to lay eggs regularly.

Optimal Lighting Conditions and Balanced Nutrition for Egg Production

For more information on lighting conditions and balanced nutrition for your egg-laying hens, you can have a look at Chapters 3, 4, and 5.

Proper light conditions and a healthy diet are both important for egg production. But a subject that is not often talked about is how you can make your eggs even tastier by feeding your chickens certain foods, and likewise, if you feed your chickens sour or bitter foods, your eggs might not have the best flavor.

Some nutrient-dense and tasty foods that can produce a great egg flavor are (Purina Mills, n.d.-d):

- Leafy greens: Greens like kale, spinach, and Swiss chard are packed with vitamins and minerals. They can contribute to richer, more flavorful yolks.

- Herbs: Herbs like basil, oregano, thyme, and parsley can impart subtle and delightful flavors to eggs. Chickens love foraging for these herbs, so if you have the space, try growing them in your yard.

- Marigolds: These colorful flowers not only add visual appeal to your garden but can also infuse a slightly spicy and aromatic note to the yolks.

- Foods rich in Omega-3: Flaxseeds and chia seeds are known for their high Omega-3 content. Including these in your chickens' diet can result in eggs with a nuttier, more complex flavor.

- Berries: Blueberries and strawberries are not only delicious but also nutritious. They can provide a sweet undertone to the yolks.

- Corn: Corn can give the yolks a slightly sweet and buttery flavor. It is a favorite treat for many chickens.

- Squash: These vegetables are not only a great source of nutrients but can also lend a rich and earthy flavor to eggs.

- Tomatoes: Tomatoes, especially the overripe ones, can add a pleasant tangy note to your eggs.

- Peppers: Hot peppers can add a spicy kick to eggs, while sweet peppers can contribute a mild and sweet flavor.

- Worms and insects: While bugs may not sound appealing to you, they certainly are to your chickens, which is a good thing since they are full of protein. Free-range chickens that eat loads of them produce a rich eggy flavor.

Techniques for Collecting and Storing Eggs Safely

Egg collecting and storing is a very simple process once you get into a good routine. While it might seem intimidating at the start, remember this, we all break a few eggs once in a while, so do not worry about having the perfect system right off the bat. It will take time to figure out what works best for you from collecting, cleaning—or not cleaning—to storing and labeling.

Proper Egg Collection

Once your chickens begin laying you will have to collect and properly store the eggs every day. Most chickens tend to lay in the morning but the exact time that your flock will choose to lay their eggs will be unique. Some keepers haul in their eggs at 6 a.m. sharp, others have to wait until 10 a.m. Find out when your chickens lay most of their eggs and go from there.

Collect the eggs as soon as your hens lay them; this ensures that you are getting the eggs while they are still warm and super fresh and reduces the risk of any spoilage or accidental cracks.

You will mostly find the eggs in your nesting boxes and occasionally in odd corners of the coop or even outside. If a hen is still laying on her nest you can gently reach your hand under her to see if an egg is there and grab it. If no egg is present, check back with that hen in a few hours. If a hen protests when you try to reach under her, have no fear, simply pick her up confidently and remove her from the nesting box to collect the eggs.

Handling Eggs With Care to Prevent Cracks and Contamination

A bit of tender care goes a long way when collecting eggs from your backyard chicken coop. Remember to treat each egg like the treasure it is, cradling it gently to prevent any mishaps. Bring some old rags or tea towels with you to provide a buffer when filling your egg basket—too many eggs jostling against each other can lead to cracks.

Cleaning and Sanitizing Eggs, If Necessary

In Chapter 1 we learned about the age-old debate of whether or not eggs should be washed. When it comes to backyard chickens the only thing that matters is your personal comfort level. No matter your stance in this debate, the most important thing is to ensure proper care and cleanliness of newly collected eggs.

If you encounter an egg with some dirt or debris on it, there is no need to worry. It is usually best to give it a gentle cleaning using a dry cloth or a fine brush. But here is the deal: Avoid using water unless absolutely necessary. Washing your eggs strips away the natural protective 'bloom', which acts like a shield against bacteria. If you do wash your eggs in water, make sure it is lukewarm to prevent the egg from contracting or expanding suddenly, which can cause issues like breakage or cracks. After washing, dry your eggs thoroughly before putting them away in the refrigerator to prevent bacteria from penetrating the vulnerable egg.

Frequency of Egg Collection to Maintain Quality

How often should you pop over to your coop for egg duty? Chickens typically lay their eggs in the morning, so making it a daily habit around the time that your flock lays their eggs is pretty smart. This not only keeps your eggs super fresh but also minimizes the chances of any accidental breakage or spoilage. It is also a good idea to keep a record of your egg collection, as it can help you understand your hens' laying patterns and catch any irregularities. Lastly, remember to check your coop or run for any hidden nests. Some sneaky hens might prefer laying eggs in out-of-the-way spots, so a thorough search can avoid any surprises and ensure all your lovely eggs are gathered in good time.

Egg Grading and Quality Assessment

If you plan on keeping the eggs you have collected for yourself and some select friends then you do not have to worry at all about egg grading and quality assessments. You will only need to consider this if you plan to have a large farm and distribute your eggs to grocery stores and restaurants and are looking to get a USDA grade mark.

The USDA, specifically the Livestock and Poultry Program's Quality Assessment Division, offers a voluntary program to officially grade eggs based on established quality and size standards. USDA graders oversee the grading process, ensuring compliance with these standards. Eggs meeting the criteria receive the USDA grade mark and are categorized as U.S. Grade AA, A, or B. This grading system helps standardize the assessment of egg quality, making it easier for buyers, sellers, and consumers to communicate about the eggs' attributes. It is important to note that only eggs processed under USDA grader supervision are eligible for this certification (U.S. Department of Agriculture, 2015).

Creative Ideas for Using Surplus Eggs

Having fresh eggs every day is a blessing, but what happens if you do not use them all in a week or even two weeks? Eggs can stay fresh for months and it would be wasteful to throw them away.

If you cannot eat them, give them away, or bake enough with them, your best solution might be to freeze them.

Freezing eggs can be a practical solution if you have an abundance of eggs that you would like to preserve for later use. However, there are some important considerations to bear in mind when freezing eggs.

Firstly, it is important not to freeze eggs in their shells. The liquid inside can expand during freezing, causing the shells to crack. Instead, separate the egg whites from the yolks. Place the egg whites in an airtight container or an ice cube tray and then freeze them.

Once frozen, transfer the individual portions to a freezer bag for extended storage. If you have leftover egg yolks, consider freezing them as well. To prevent the yolks from developing a gel-like consistency when thawed, add a pinch of salt or sugar before freezing. Store them in airtight containers or ice cube trays, then transfer them to a freezer bag.

Alternatively, you can freeze whole eggs by gently beating them together before freezing. Crack the eggs into a bowl, whisk the yolks and whites to combine, and then pour the mixture into an ice cube tray or an airtight container. This method is particularly effective if you intend to use the eggs for baking or cooking.

Using freezer-safe containers is imperative to prevent freezer burn and the absorption of odors that could impact the quality of the frozen eggs. When you are ready to use the frozen eggs, it is best to thaw them in the refrigerator and use them within 24 hours. Keep in mind that frozen eggs may undergo slight texture changes after thawing, making them best suited for dishes like baking, casseroles, and omelets.

A few other great ideas to use up your extra eggs are (Winger, 2016):

- Classic scrambled eggs: Whip up fluffy scrambled eggs with a pinch of salt and pepper.

- Omelets: Create personalized omelets with your favorite fillings like cheese, veggies, or ham.

- Fried eggs: Cook sunny-side-up or over-easy eggs for a quick and delicious breakfast.

- Egg sandwich: Make a breakfast sandwich with eggs, bacon or sausage, and cheese.

- Hard-boiled eggs: Prepare hard-boiled eggs for snacks or to top salads.

- Egg salad: Mix chopped hard-boiled eggs with mayonnaise and seasonings for a tasty sandwich filling.

- Deviled eggs: Scoop out yolks, mix with mayonnaise and mustard, and fill egg whites for a classic appetizer.

- Quiche: Bake a savory quiche with eggs, cream, cheese, and various fillings.

- Frittata: Cook a skillet frittata with eggs and a variety of veggies and meats.

- Egg drop soup: Add beaten eggs to hot broth for a comforting Chinese-style soup.

- Custard: Whip up a creamy custard dessert with eggs, milk, and sugar.

- French toast: Dip bread in an egg and milk mixture, then fry for a sweet breakfast treat.

- Homemade pasta: Mix eggs and flour to create fresh pasta dough.

- Crepes: Make thin, delicate crepes using eggs, flour, milk, and a bit of sugar.

- Egg wash: Brush beaten eggs on pastries and bread for a golden, shiny finish.

- Bread pudding: Bake a rich bread pudding dessert with eggs, milk, and spices.

- Egg fried rice: Whip up a quick and flavorful fried rice dish with scrambled eggs.

- Egg curry: Prepare a spicy egg curry with boiled eggs in a flavorful sauce.

- Egg salad sandwich: Create a creamy egg salad for a satisfying sandwich.

- Hollandaise sauce: Whip up a classic hollandaise sauce to serve with Eggs Benedict.

- Mayonnaise: Combine eggs, oil, vinegar, and mustard to make homemade mayo.

- Lemon curd: Make a tangy lemon curd with eggs, lemon juice, sugar, and butter.

- Meringue: Beat egg whites and sugar to create light and fluffy meringue for pies.

- Homemade eggnog: Mix eggs, cream, sugar, and spices for festive holiday eggnog.

- Egg wash for baking: Use an egg wash to give baked goods a shiny, golden crust.

- Egg white face mask: Create a DIY face mask with egg whites for an at-home skin tightening treatment.

- Hair conditioner: Use egg yolks as a natural hair conditioner for added shine.

- Gardening: Crush eggshells to add calcium to garden soil for healthier plants.

- Crafts: Decorate blown-out eggshells for Easter or as ornaments.

- Science experiments: Use eggshells to demonstrate various scientific principles.

Safe Egg Storage

Eggs need to be stored properly to maintain their freshness. It is a good practice to place them in egg cartons or other egg-safe containers, ensuring each carton is labeled with the collection date. Make sure to rotate your eggs by placing the newer ones at the back of the carton or on the bottom of the container stack.

If your eggs have their bloom intact you can store them at room temperature with no worries for a few weeks at a time. However, if you want to extend the shelf life of your eggs, the refrigerator is your best bet. If you have washed your eggs, they must be kept in the fridge.

Imagine you just came home from a long vacation and you open your fridge to see your eggs sitting happily in their container, but something is missing: the collection date. Who knows how long that egg has been sitting there?

This happens to the best of us. Luckily, there is an easy solution called 'the float test'. To check if your egg(s) is still fresh, gently place it into a deep bowl of cold water. If it floats, it has too much of an air pocket inside, indicating it might have gone bad and the best place for it is on the compost pile. Any eggs that sink or drift towards the bottom are still viable to use. If an egg sinks on its side, it is still fresh, while, if it sinks on its end, or with one end pointing up, it is slightly older but still good to eat.

Alternatively, you can do the "candling" test. Simply shine a small but very bright light up against the eggshell. If your egg has gone bad, you will see a large air pocket. Compare the egg against a fresh egg and an egg that is a few weeks older so you can tell the difference.

Temperature and Humidity Considerations

To store backyard eggs properly, it is advisable to use a refrigerator set at temperatures between 45°F and 50°F (Arcuri, 2023). This temperature range helps to preserve the freshness and quality of the eggs. The humidity level inside the refrigerator should be maintained at around 70% to 80%. Achieving this humidity level can be as simple as placing a small bowl of water in the fridge, which helps prevent the eggs from drying out and maintain their ideal storage conditions. These measures ensure that your freshly laid backyard eggs stay in optimal condition until you are ready to use them.

Choosing the Right Storage Containers

The optimal storage containers for backyard chicken eggs will safeguard them from breakage, moisture loss, and contamination. There are several viable options to consider:

- Egg cartons: These conventional holders, available in cardboard or foam, are a favored choice. Purpose-built to cradle eggs securely, they provide robust protection against damage. You can procure cartons in various sizes to accommodate different egg quantities.

- Plastic egg crates: Renowned for their durability and ease of cleaning, plastic egg crates come in various sizes and are reusable. Many feature a locking mechanism to ensure egg security.

- Egg skelters or racks: These devices are designed to facilitate the use of the oldest eggs first, helping you maintain freshness. Typically constructed from metal or plastic, they offer clear visibility of your egg collection.

- Egg baskets: If you prefer a more rustic or aesthetically pleasing option, consider using egg baskets crafted from wire or wicker. While they may not offer the same level of protection as cartons or crates, they can add a touch of charm to your storage.

- Egg storage containers: Specially designed containers with individual compartments are available, preventing egg breakage and enabling you to utilize the oldest eggs first.

- Refrigerator trays: Designed to fit snugly inside your fridge, these trays securely hold eggs while optimizing space and helping maintain a consistent temperature.

Remember to store your eggs with the pointed end facing downward, as this preserves the air sac and keeps the yolk centered. It is also prudent to label your egg containers with the collection date to help you track their level of freshness easily. By selecting appropriate storage containers, you can effectively extend the shelf life of your backyard chicken eggs while preserving their quality.

Interactive Exercise

Create an egg production log to track the number of eggs laid by each hen in your flock. Monitor and analyze the data over time to identify patterns and trends. Use this information to fine-tune your management practices and make adjustments to maximize egg production.

Chapter 9:

Raising Chickens for Meat

In this chapter, we will explore the process of raising chickens specifically for meat production. In order to understand this process properly, we will learn how to select the right breeds, provide proper care and nutrition, and follow humane and safe practices for harvesting and processing chicken meat.

Differences Between Layers and Meat Breeds

In a balanced poultry farm, both layer and meat breeds have their roles to play. Layer breeds keep the egg supply steady, while meat breeds ensure a bountiful harvest of delicious poultry. Whether you are inclined towards egg production or meat, each type of breed has a unique contribution to make to your flock. Understanding these differences will help you make the right decisions when selecting birds for your backyard flock.

Layer breeds are the top egg producers. They are known for their consistent and abundant egg-laying capabilities but few praises are given for their meat quality. Layers are typically smaller in size and are favored by backyard farmers.

Meat breeds, on the other hand, are raised primarily for their meat. They are the heavyweight champions of the poultry world. While they may not lay as many eggs as layer breeds, they excel in providing quality poultry for the dinner table.

Breeds Suitable for Meat Production

Chicken breeds well-suited for backyard broiler production have often been selectively developed for their meat-producing qualities, including rapid growth and good meat-to-bone ratios. Here are some popular backyard broiler chicken breeds (Morning Chores, 2019):

- Cornish Cross: This breed is one of the most common choices for backyard broilers and the commercial poultry industry. They grow rapidly and produce high-quality meat. However, they can be prone to health issues due to their rapid growth, so careful management is essential.

- Red Ranger: Red Rangers are known for their excellent foraging abilities and robust health. They grow slightly slower than Cornish Cross but offer a more traditional broiler option.

- Freedom Ranger: These chickens are prized for their flavorful meat and strong foraging instincts. They are more active than Cornish Cross and take a bit longer to reach processing weight.

- Jersey Giant: Jersey Giants are one of the largest chicken breeds. While they take longer to reach maturity, they produce substantial amounts of meat with excellent flavor.

- Delaware: Delawares are a dual-purpose breed but are known for their meat quality. They have a pleasant disposition and can be a good choice for small backyard flocks.

- Barred Plymouth Rock: Often called "Barred Rocks," these chickens are another dual-purpose breed that produces both eggs and meat. They are known for their hardiness and friendly temperament.

- Orpington: Orpingtons come in various colors, including Buff, Black, and Blue. They are large birds with good meat quality and a calm disposition.

- New Hampshire Red: Similar in appearance to Rhode Island Reds, New Hampshire Reds are known for their meat production. They have a robust constitution and adapt well to backyard environments.

Remember that if you aim to have a successful broiler production it will depend not only on breed selection but also on proper nutrition, housing, and overall care. Additionally, you will need to consider your local climate and availability when choosing the best broiler breed for your backyard.

Rapid Growth and Weight Gain

There are many suitable types of broiler chickens. When selecting which breed is right for you, you will need to consider how quickly the chickens will grow as well as the ethics and sustainability of raising broilers.

Cornish Cross breeds dominate within large meat packaging industries and typically reach market weight within six to seven weeks (RSPCA, n.d.). Their lightning-speed growth rate makes them a valuable choice for both large and small-scale poultry farmers looking to raise and/or sell their own meat. However as discussed in Chapter 2, the rapid growth of the Cornish Cross puts them at a huge disadvantage physically and they tend to have a poorer quality of life, require special care and nutrition to prevent illnesses and sores, and are also born incapable of reproducing on their own.

Other broiler-suitable breeds may grow slower but they are much healthier and more sustainable as they are not sterile and their bodies are capable of holding up their weight.

Meat Quality and Flavor Considerations

In the world of backyard chicken keeping, the pursuit of exceptional meat quality and flavor becomes an art. Several key factors contribute to what makes a backyard chicken truly delicious.

The choice of chicken breed significantly influences meat quality. Particular heritage and dual-purpose breeds have gained notoriety for their outstanding meat. Breeds like the Cornish Cross and Red Broilers are celebrated for their tender and succulent meat, making them favored selections among backyard chicken keepers pursuing exquisite flavor.

Diet is paramount in shaping the flavor of backyard chicken meat. These birds are often privileged with a diverse diet, complete with foraged insects, greens, and grains. The opportunity to explore and peck at various foods imparts distinct flavors to their meat. Many backyard chicken enthusiasts choose to feed their flock with organic, non-GMO feeds, elevating the overall meat quality.

Age plays a crucial role in determining meat quality. Unlike commercial broilers—like the Cornish Cross—which are processed at a tender age, some backyard chickens—such as the Red Ranger and Jersey Giant—need more time to mature. This extended growth period allows their meat to develop richer and more complex flavors. Chickens processed when they are between 12 and 24 weeks old tend to yield meat with superior depth and character. Unfortunately, it is not advised to allow your Cornish Cross to develop past that (Talbot, 2022).

The rearing environment also plays a vital role. Backyard birds often enjoy less stress than their commercially raised counterparts, resulting in meat that is more tender and flavorful.

Finally, following proper processing methods will also aid in preserving meat quality. Employing humane slaughter techniques, coupled with meticulous chilling, is crucial. Many backyard chicken keepers prefer to manage the processing at home to ensure these essential steps are executed with the utmost care.

Differences in Care and Management

Raising chickens for meat comes with a unique set of considerations that are quite different from raising layers kept for egg production. To excel in poultry farming, a chicken keeper will need to recognize and address these distinct needs and challenges.

Ethical broiler farming starts with ensuring the birds have a good quality of life. Broilers are bred for rapid growth, which can pose welfare challenges. Therefore, they will need adequate space, access to natural light, and a clean environment. Overcrowding should be avoided to reduce stress and health problems. Responsible broiler farming also involves choosing breeds or genetic lines that prioritize both growth and the well-being of the birds.

Nutrition is another key aspect since broilers grow rapidly and have different dietary needs compared to layers. Proper nutrition and a broiler-specific diet are vital for their health, as these factors help to prevent metabolic issues and support quick muscle development without overfeeding. Ethical broiler farming involves carefully balancing their diet to prevent health problems associated with their rapid growth.

Regular health checks are a must in broiler farming and include monitoring for common problems like respiratory issues, lameness, or infections. Prompt medical attention should be given when necessary to prevent unnecessary suffering. Ethical broiler farming prioritizes the well-being of the birds throughout their relatively short lives.

Broilers also benefit from opportunities for natural behaviors such as dust bathing, perching, and foraging. Providing environmental enrichment like straw bales or hanging objects can reduce stress and boredom, contributing to their overall welfare.

Feed Requirements and Nutrition

Just like layers, broilers require feed that has been specifically developed for them. Some broilers are dual-purpose, in which case, you will want to adjust their feed accordingly. To learn more about nutritional requirements for meat birds, you can go back to Chapter 4.

Space and Housing Considerations

When it comes to raising broiler chickens, providing the right space and housing is paramount to their well-being and growth. Their living conditions will directly impact their health and the quality of their meat.

Broilers need adequate space where they can move around, exercise, and have clear access to food and water. Typically, each broiler chicken requires around 0.8 to 1 square foot of indoor space in a controlled environment (University of Minnesota Extension, n.d.). However, in more extensive outdoor systems, they might need 2 to 3 square feet each. Failure to provide the necessary space can cause overcrowding, which leads to stress, disease, and poor meat quality.

Broiler houses should be well-ventilated to ensure good air quality and temperature control. Proper ventilation helps to reduce the unhealthy buildup of gasses from droppings, which can harm the birds' respiratory health. Adequate insulation and heating may be necessary in colder climates to maintain a comfortable temperature.

The flooring should be designed to prevent leg problems, which is a common issue in broilers. A good option is to provide litter, like wood shavings or straw, that helps support their leg health and keeps the house clean.

Good lighting is also crucial to regulate the birds' activity and feeding patterns. Natural light or controlled artificial lighting can influence their growth rates and overall health.

Raising and Caring for Meat Chickens

Brooding Meat Chicks

Brooding meat chicks is the essential first step in raising poultry for meat production. During this phase, you have to provide the chicks with a warm and comfortable environment. The brooder, their initial home, needs to maintain a temperature of around 95°F (35°C) during the first week, gradually decreasing by 5°F (2.5°C) each week until the chicks are fully feathered (USDA APHIS, 2022). To achieve this, you can use heat lamps or brooder heaters, but make sure to create a cooler area within the brooder where chicks can retreat if they get too warm.

You will also need to select the right bedding material for your brooding chicks. Pine shavings or straw are excellent choices because they offer comfort and absorbency, making cleaning easier.

Nutrition is of utmost importance in the early stages of a meat chick's life. Choose a high-quality starter feed with a protein content of about 20-22% designed specifically for meat chicks (USDA APHIS, 2022). Additionally, ensure they have constant access to clean, fresh water.

Regularly observe the chicks during this time and look out for any signs of distress or illness. Constant vigilance allows you to intervene swiftly and prevent any potential problems. Since chicks are social animals, they will need companionship, but be careful not to overcrowd them, as it can lead to stress and health issues.

Ensure that your brooder has proper lighting, with approximately 18 hours of light daily for the first few weeks to encourage healthy eating habits and growth. Adequate ventilation is also necessary to maintain fresh air and reduce ammonia buildup from droppings.

If your brooder is outdoors, make sure it is predator-proof to protect the vulnerable chicks. As the chicks grow and feather out, gradually introduce them to the outdoor environment to help them acclimate to changing temperatures.

In essence, brooding meat chicks is a delicate process that lays the foundation for raising healthy and thriving birds, ensuring a future supply of robust, flavorful meat. If you follow these essentials of brooding meat chicks from the start, you can ensure that they have a good start and a better chance at a healthy level of growth.

Temperature and Humidity Control

The brooder, their initial home, needs to maintain a temperature of around 95°F (35°C) during the first week, gradually decreasing by 5°F (2.5°C) each week until the chicks are fully feathered. To achieve this, you can use heat lamps or brooder heaters, but it is crucial to create a cooler area within the brooder where chicks can retreat if they get too warm.

Maintaining the right humidity in your brooder is vital for the well-being of your meat chicks. Especially during the first few weeks of their lives when they are most vulnerable (USDA APHIS, 2022).

Week 1: Aim for a humidity level of around 50-60%. This higher humidity helps prevent the chicks from getting dehydrated.

Week 2-3: Gradually decrease the humidity to around 40-50%. This mimics a more natural environment as they start to grow feathers.

Week 4 and beyond: Reduce the humidity further to about 30-40%. By this time, your meat chicks should have developed enough feathers to regulate their body temperature better.

Remember, these are general guidelines, and it is essential to monitor your brooder's humidity levels regularly. Too much humidity can lead to respiratory issues and mold growth, while too little can cause dehydration and other health problems.

Use a hygrometer (a humidity-measuring device) to keep tabs on humidity. Adjust the humidity by adding or removing water from your brooder setup, such as the bedding material. Keeping humidity in check contributes to your meat chicks' overall health and ensures a smooth transition as they grow into hearty birds.

Proper Feeding and Nutrition

Broiler chicks will need high-energy foods throughout their growth cycle. To learn more about how to properly feed your broilers refer back to Chapter 4.

Disease Prevention and Management

The best way to prevent diseases in your chicks is to make sure that the eggs you purchase come from a reputable hatchery with high health standards. If you are brooding a clutch from your flock, make sure that the mama hens are free of disease and harmful parasites.

Managing Growth and Weight Gain

Rapid and uncontrollable weight gain is a common issue among broilers. To avoid or mitigate the potential health complications of your chicks putting on too much weight before their little bodies are strong enough for it, carefully portion their food and place them in a healthy feeding schedule. Monitor their weight regularly and make sure that they are growing at a healthy pace.

To learn more about health risks associated with rapid weight gain and what feed is best for broiler chicks and pullets, take a look at Chapters 4 and 5.

Harvesting Chicken Meat Safely and Humanely

Whether you raise chickens for eggs only, meat only, or a combination of the two, there comes a time in most keepers' lives when it becomes necessary to cull a chicken. Sometimes it is a way of humanely ending the suffering of a sick or injured chicken, or because a hen has exceeded its peak laying potential, or because the chicken was a rooster. Even keepers who view their chickens as cherished pets and companions encounter situations that require the euthanizing of their chickens.

Regardless of your motives behind keeping chickens, an unfortunate truth is that death is a natural part of everything, from plants to animals. If you want to provide yourself and your family with fresh, healthy, and sustainable chicken meat from your backyard, then humane slaughtering is a necessity. In this section, we will cover some basics of culling the flock.

If you are new to the butchering process, it is important to be aware that the chicken's body will likely exhibit some involuntary movements. This might be unsettling if you are unprepared or have never experienced the chicken slaughtering process.

Even though the chicken is making these movements, it is important to understand that it is not in any pain and is truly dead. Involuntary movements may include twitching and strong convulsions and can persist for a minute or two.

While this experience may be shocking at first, know that these post-mortem movements are entirely normal. Movements may also stop suddenly and then start again without warning, so it is good to be prepared for any sudden changes. Even if the chicken opens its eyes after slaughtering, it does not mean that the chicken is alive.

These involuntary movements are not unique to any one slaughter method. Even if your chicken has been fully decapitated it may still make a lot of movement. Again, your chicken is not in any pain because it has been fully beheaded and is not living. This is completely normal. Everything is okay.

However, if you are in any doubt that your chicken is not dead after severing its carotid artery, cut their neck again, going deeper than you did before.

Humane Slaughter Practices

The two most important things to consider when slaughtering chickens are humane treatment and sanitary practices.

When slaughtering chickens, make sure that your knife, ax, or hatchet has the sharpest blade you have ever held in your life. The sharper the edge, the more humane and painless the process will be. Always sharpen your blade between every few birds. You will know your blade is sharp enough if the lightest pressure will slice your skin. Do not attempt to slaughter a chicken if your blade is not sharp enough.

When you make the cut, do so as quickly and firmly as possible. The faster, deeper, and harder you can make your cut the more humane the process will be (Wyss, 2017).

The use of culling cones is recommended and is one of the most humane tools for slaughtering birds (Lobermeier, n.d.). Not only do they give you better control over your cut, but culling cones help to ensure thorough blood drainage and prevent any bottlenecks in this critical stage.

To slaughter a chicken humanely and with care, begin by holding the bird with their wings at their sides. Then 'swirl' the birds in circles in the air to disorient them. Quickly flip the bird upside-down and place their head through the end of the cone. As soon as you do so, hold the head and swiftly and firmly cut the neck at the jugular just below the head and above the trachea. Make sure your knife or blade is extremely sharp—so you do not want to have to cut twice. Cut deep enough to fully sever the carotid arteries. Leave the slaughtered chicken upside down until they have bled out for at least two minutes.

Properly bleeding your meat is critical. Failing to do so leads to an excess of blood remaining in the meat, which leads to bad-tasting meat and a shorter shelf life. However, the biggest concern is that the residual blood can lead to the growth of harmful pathogens—regardless of how your meat is processed or how quickly it is chilled (Nakyinsige et al., 2014).

Once you have completed the slaughtering phase, the next step is to thoroughly clean the chicken with a strong water hose. Chickens are frequently coated with dirt and feces, so washing the chicken ensures that the meat is clean and ready for processing. So, do not skip this step.

Choosing the Right Slaughtering Method for Your Comfort and Skills

There are various ways to slaughter a chicken. Some are more practical for small-scale farmers and chicken keepers, while others are better suited for big-scale industries. However, when it comes to ending the life of a living creature, there is no easy way to do it, so it is best to be well-informed and prepared.

The choice of slaughter method often depends on personal preference, cultural practices, and the resources available. Choose a method that aligns with your values and your comfort level with the process. Additionally, ensure that whatever method you choose is done humanely and quickly to minimize any suffering.

The three most common types of slaughtering for backyard, homestead, and small farm chicken keepers are:

- cutting the carotid arteries

- decapitation

- neck-snapping (this method is not recommended for humane slaughter, as it does not always work correctly the first time and causes unnecessary harm).

Cutting the neck is one of the most popular choices during butchering season. It requires a very sharp knife and you can either have someone hold the chicken for you while you cut their neck, or use a kill cone.

Kill cones can be easily attached to a sturdy structure like a tree or a fence post. The idea is to create a calm and controlled environment for the chickens. Kill cones also provide you with the most control over your blade and do not require the aid of a second person.

However, a common slaughtering practice is to simply hang the bird upside down from a sturdy branch or a reliable rafter. This requires two people, one to hold the chicken still and its wings in place and another to slice the bird's jugular. Make sure to hang the bird around at chest height (Linden, 2015). This ensures you have the most control over your blade, keeps you safe, and prevents you from having to run inside to wash your face or mouth.

Decapitation can be done in several ways but usually requires two people per chicken. While one person holds the bird still, the second person simply chops the chicken's head off with an extremely sharp hatchet or an ax.

Techniques to Minimize Pain and Stress

Before culling the flock, make sure to limit your chickens' food intake for the day. Dispatching a chicken with a full crop, stomach, and intestines can be quite messy and can also make the slaughtering process more stressful for the chicken (Lobermeier, n.d.).

Legal and Regulatory Considerations for Slaughter

Cities and urban areas often enforce regulations that determine the number of chickens you are allowed to keep as well as their spatial requirements and the permissibility of roosters. Nevertheless, explicit laws regarding slaughter are often absent. There is a growing recognition that urban areas must address multifaceted issues associated with urban agriculture, including slaughter. At present, the laws have not kept pace with the proliferation of urban homesteading; consequently, engaging in the processing of backyard chickens typically does not entail legal violations (Ploetz, 2013).

However, bear in mind that selling backyard processed meat to others constitutes a violation of the law unless you have undergone a USDA inspection. This option remains viable solely when the meat is designated for exclusive consumption by your family.

In certain U.S. states, it is lawful to sell home-butchered chickens directly from the farm on the same day as slaughter. However, it remains illegal across the board to slaughter chickens at home with the intent of selling them elsewhere. This endeavor necessitates additional licensing, with specific requisites varying from state to state. It is advisable, therefore, to consult your local extension office to ensure compliance with pertinent regulations before embarking on the sale of home-raised meat (Ploetz, 2013).

Processing Chicken Meat

As a conscientious poultry enthusiast, you will likely find yourself in situations where you need to process chickens, be it for securing your family's food source or tending to your flock's welfare. In this guide, we explore the key steps and tips for processing chicken meat at home. It is all about making the most of your backyard poultry setup while keeping things down-to-earth and humane.

Feather Removal and Cleaning

The first step in chicken processing requires a large stockpot filled with vigorously boiling water—between 125-132° F—adequate to fully immerse the bird. You can either use a propane setup or even do things the old-fashioned way, over an open campfire. If you do decide to use the campfire, make sure that you monitor the temperature of your water carefully.

With care, take hold of the chicken's feet and immerse it in the boiling water up to the point where the feathers end—typically at the first leg joint. There is no need to dunk the scaly feet. Hold it there briefly, for about two seconds, and then lift it out, allowing excess water to drain off.

Now, move the wet chicken to the area where you will be removing its feathers. The hot water dunking makes feather removal more manageable. To pull out the feathers effectively, work against their natural direction. Chickens are usually cooperative in this regard, though it can be a meticulous process. For other poultry, like ducks, feather removal can be more challenging.

Pay special attention to areas where feathers tend to be stubborn—the wingtips, tail, and around the vent (rectum). Do your best to remove all the small down feathers until the chicken is completely feather-free. If you miss a few feathers in the heat of the moment, do not worry. You can always pluck them later before cooking. Persistence is key in this task, especially when more chickens are waiting their turn.

Evisceration and Butchering Techniques

Although the term 'evisceration' might appear somewhat intimidating, in this context, it means the removal of the internal organs from the chicken, essentially preparing it in a similar way to the poultry you would purchase at a store. This step, while somewhat challenging, plays a vital role in the process of dispatching a bird. At this stage, the primary concern revolves around preventing any contamination of the meat by the chicken's excrement, which can be tricky.

The first step involves removing the chicken's feet with precision. You will need to make precise cuts at the junction between the leg and foot joints. By using a well-sharpened knife, you can cleanly sever the connective tissue without affecting the underlying bone structure, resulting in a meticulous detachment of the foot from the leg.

It is worth mentioning that the chicken's extremities are completely usable and should be kept. After removing the toes, go through a thorough cleaning process, and then store them in the freezer for later use in making a rich and gelatinous broth. Setting aside a designated space for these chicken feet for this specific purpose is crucial.

The next step is to remove the internal organs and neck. Start by detaching the head high on the neck—if it is not detached already—and the legs where the feathered skin starts. Rinse the carcass thoroughly. Cut off the neck near the shoulders, followed by a midline cut between the breastbone and the tail. Create a circular cut around the bird's vent, being careful not to cut into the intestines.

Upon reaching the top of the chest cavity, encircle your fingers around the organs and gently pull them out through the abdominal opening. Exercise caution during this step and make sure not to rupture the gallbladder, which contains dark green fluid that can negatively affect the taste of the meat.

Thoroughly clean all edible parts of the chicken. Minor fecal contamination can be washed off with water. Remember, the liver, heart, gizzards, as well as the neck and feet, should never go to waste. They can all be utilized for nutritious purposes, ensuring that every part of the bird contributes to nourishing you and your loved ones. Store your meat in the refrigerator, after 8-12 hours you can freeze the meat for later use (Linden, 2015).

Do not freeze your chickens until they have been adequately chilled to a temperature of 40°F (4°C) or below (University of Minnesota Extension, n.d.). Attempting to freeze a large number of unfrozen birds simultaneously in a typical home freezer can overload the appliance, resulting in slow freezing of the unfrozen birds and partial thawing of items that were already frozen. To ensure proper freezing and maintain food safety, freeze a small number of unfrozen birds at a time in your home freezer.

Packaging and Storage Guidelines

After completing the processing phase, it is time to prepare it for long-term storage. Gallon-sized freezer bags or even a vacuum food sealer are good options for preservation. Label your bag with the date your chicken was packaged and the type of cut or organs inside. Then, promptly transfer them to the freezer.

You can freeze the chicken whole with the saved organs inside the cavity, similar to what is done in supermarkets. Alternatively, freezing the organs separately is another option. If you prefer smaller portions, consider cutting the chicken into eight pieces. For those interested in culinary experimentation, grinding the chicken for sausage before freezing is also a possibility. The method you choose offers flexibility in how the chicken is handled for storage.

Food Safety Considerations

There are various perspectives on the topic of self-processing meat, particularly when it comes to chickens. On one side, there is a valid concern for public health and the important role played by regulatory agencies like the USDA and FDA in ensuring food safety. Their strict regulations are essential to prevent illnesses, foodborne diseases, and contamination.

However, sometimes, the emphasis on safety can discourage individuals from learning the skills necessary for self-sufficiency in raising and processing their meat. This transition, especially for those moving from urban to more self-reliant lifestyles, can raise questions about sanitation risks and the possibility of contamination, like salmonella.

The truth is, with a basic understanding of food safety practices, chicken butchering, and processing can be done safely. Many educational resources are available and created to empower people to take control of their food production while maintaining high standards of hygiene and safety.

Some basic food safety for culling and butchering your chickens include:

- washing your hands

- washing your poultry to ensure there are no fecal contaminants

- having a sanitary workspace

- fully bleaching and sanitizing all work equipment after the slaughtering and butchering process

- not feeding chickens 6-8 hours before slaughtering

- wearing protective gloves to avoid cutting yourself

- making sure that the chickens you slaughter are disease-free, well-developed, and healthy

- cooling your poultry immediately after processing

- storing chicken at the appropriate temperature (Hamre, n.d.)

Safe Storage and Shelf-Life Guidelines

Before cooking your freshly slaughtered chicken, allow it to rest in the refrigerator for 12-24 hours to allow any tension and excess adrenaline in the meat to dissipate. If you freeze your chicken only a few hours after slaughtering, allow it to thaw in the fridge and rest for a day or two before cooking. This way, you can avoid any possibility of having tough meat.

When you are ready to use your frozen meat, allow it to thaw in the refrigerator in water. Do not refreeze thawed chicken. Chicken stored in the fridge should be kept at 40F (Food Safety and Inspection Service, 2013).

Interactive Exercise

Calculate the cost of raising chickens for meat by considering factors such as feed expenses, housing costs, and processing equipment. This exercise will help you understand the financial aspects and potential savings of raising your own meat chickens while ensuring quality and sustainability.

Chapter 10:

Dealing With Predators

A primary concern for all chicken keepers, whether they reside in urban, suburban, or rural environments, is securing their chicken coop against potential predators. Predators like foxes and snakes can pose constant threats to the safety of your chickens, necessitating a range of protective measures. In this comprehensive chapter, we will explore the various aspects of predator prevention for chicken coops, with a primary focus on safeguarding the well-being, health, and peace of mind of your flock. We will begin by identifying the types of chicken coop predators commonly found in different geographic regions and then discuss effective strategies for mitigating these threats

Identifying Common Predators of Chickens

In the world of poultry keeping, knowledge is power, and knowing your flock's potential adversaries is essential for their protection. This comprehensive list delves into the various common predators that can threaten chickens. From the crafty raccoon to the elusive fox, understanding these potential threats is the first step in ensuring the safety and security of your feathered companions.

Common Predators

Although we discussed a few common coop predators in Chapter 2 below is a more complete and comprehensive list. Explore this guide to learn more about these chicken foes and how to shield your flock from harm.

Raccoons

Raccoons are infamous for their chicken-hunting skills. These highly adaptable creatures skillfully infiltrate coops, often targeting both eggs and vulnerable birds. They are highly persistent and tend to engage in overkilling, usually leaving more deceased birds than they can immediately consume. Indications of a raccoon intrusion could include missing chickens, often accompanied by scattered feathers, with eggs remaining untouched.

Foxes

Foxes are prevalent in various regions across America and are known for their cunning resourcefulness and their unwavering pursuit of poultry. These sly predators might leave behind signs like the absence of chickens and the presence of scattered feathers.

Coyotes

Widespread across the United States, coyotes are renowned for their proclivity to attack chickens, particularly in suburban and rural locales. Exercise vigilance and remain on the lookout for missing birds, as this could be a potential indicator of coyote involvement.

Snakes

Various snake species, such as rat snakes and racers, are irresistibly drawn toward chicken coops, primarily enticed by the prospect of eggs and rodents. Since their primary focus tends not to gravitate toward adult chickens, their presence might often elude immediate detection but still poses a formidable risk. Be on the lookout for discarded snake skins, or consider employing guineas as a deterrent against snake threats.

Opossums

Opossums, opportunistic scavengers by nature, are a particular threat to young chickens. They have cultivated a notorious reputation as nest robbers. Damp or bloody feathers or a disheveled nest bereft of eggs could serve as indicators of opossum involvement.

Hawks and Owls

Skilled hunters within the avian hierarchy, hawks, owls, and their ilk pose credible threats to chickens, particularly when these birds roam freely. Exercise heightened vigilance during both daylight and nighttime hours, as these birds of prey pose a potential threat at all hours.

Skunks

While not continuously featured on the roster of chicken predators, when skunks do strike, they unleash a full-scale assault. These omnivorous creatures may resort to egg pilfering or target small chickens when granted access to the coop. Look out for potential disturbances around the coop and the inexplicable disappearance of eggs.

Weasels

Weasels, characterized by their diminutive stature and nimbleness, frequently engage in acts of wanton killing. They specialize in infiltrating confined spaces and can wreak havoc within flocks. Signs of weasel attacks include intact yet lifeless birds or chickens bearing pulled-out intestines.

Feral Cats

Ownerless and feral cats may harbor proclivities that pose risks to chickens, particularly to fledgling chicks. These stray felines, possessing adept hunting skills, often set their sights on chicks, leaving minimal evidence in their wake, save for the conspicuous absence of birds.

Dogs

Even domestic canines, when devoid of prior exposure to chickens, can cultivate predatory instincts. Maintain a watchful eye for instances where chickens are entirely missing, for dogs that are not accustomed to these birds can potentially abduct and carry them away.

Bobcats

In select parts of America, bobcats can be formidable threats, particularly in rural settings. Their techniques involve targeting the jugular or head during attacks. Be on the lookout for scratch marks surrounding attacked birds, as bobcats, known for their propensity to camouflage their prey, might display these patterns.

Black Bears

Black bears, in particular, possess an affinity for chicken coops, largely driven by the prospect of securing an effortless meal. These sizable mammals boast an acute sense of smell capable of detecting the scent of chickens, their feed, and eggs from considerable distances. The presence of black bears often results in a disheveled scene, featuring carcasses, remnants, and a distinctive odor.

Mice and Other Rodents

The radar for potential threats should extend to encompass smaller yet still troublesome predators. Indications of their presence include the conspicuous absence of eggs, feeders, and waterers displaying signs of nibbling, the presence of diminutive pellet-like droppings, and the unexpected appearance of nests within the coop.

Humans

Regrettably, humans might occasionally assume the role of perpetrators. In cases where suspicions of foul play arise, you may have to uphold stringent security measures, possibly including the deployment of a guard dog for protection.

Recognizing Predator Signs

When you are raising chickens, protecting them from potential threats is a top priority. One crucial aspect of safeguarding your flock is learning how to recognize signs of predator activity and identifying the culprits behind the attacks on your chickens so that you can take proactive measures to keep your chickens safe.

Footprints, Droppings, and Other Physical Evidence

Predators often leave behind telltale signs of their presence. Keep an eye out for footprints around your coop and chicken run since predators have distinct paw or claw prints that can help you identify them. For instance, raccoon prints typically resemble miniature human handprints, while fox prints are more dog-like but with prominent claws.

Examining droppings is another way to discern the type of predator you are dealing with. Raccoon scat, for instance, often contains undigested food remnants like berries or seeds. Fox droppings are tubular and tapered at the ends, resembling a small dog's feces. Snakes, on the other hand, usually leave feces filled with bones and fur (Freeman, 2022).

Damage Patterns on Coop and Fencing

Inspecting your coop and its surrounding fencing can provide valuable clues. Raccoons, known for their dexterity, often leave evidence of tampering with latches or locks. They may even peel back wire mesh to access your chickens. Foxes are skilled diggers and might create burrows underneath your coop or run.

Birds of prey like hawks and owls typically leave evidence of their attacks on your chickens. Look for feathers scattered around the coop area, especially if you notice injuries on your birds consistent with talon marks. Snakes may enter through small openings or gaps in the coop, leaving a clear path of entry.

Observing Predator Behavior

Sometimes, the best way to identify a predator is by observing its behavior. If you witness a predator stalking your chickens, take note of its physical characteristics, such as size, fur or feather coloration, and any distinctive markings. These observations can help you confirm its identity.

Deterrence and Prevention Strategies

When it comes to safeguarding your cherished flock from potential threats, a proactive approach is essential. The following strategies will help you fortify your chicken coop against cunning predators using both deterrence and prevention methods.

Make sure to vary your protective strategies. Your arsenal for ensuring your chickens' safety and well-being can include fortifying your coop's construction, incorporating nighttime protective measures, employing natural deterrents, and following safe free-ranging practices.

Predators are clever and can become accustomed to fixed routines and objects. For example, if you have a scarecrow in your yard, move it to different positions from time to time to keep predators guessing (Smith, 2020e).

For more information on predator-proofing your coop refer back to Chapter 2 in the section "Tips for Building a Predator-Proof Coop".

Secure Coop Design

In Chapter 2, we discussed how to build a predator-proof coop, below is a brief overview of some of the information covered in that chapter and a more in-depth look at predator deterrents.

Make sure to use hardware cloth with tiny 1/2-inch openings to seal up any gaps in your coop. This material is incredibly strong, unlike chicken wire, which is more suitable for arts and crafts. Ensure that you bury the hardware cloth at least six inches deep in the ground around your whole coop and extend it out about a foot from your coop's edges. This trick will stop any digging predators from finding an opening or burrowing their way in.

Do not leave any holes unsealed. If you have weasels causing trouble, check for tunnels they might be using, since they typically love sneaking in through the ones made by rodents.

Even if your coop windows have screens, add some hardware cloth. Screens are good for keeping bugs out, but hardware cloth is your guardian against unwelcome visitors.

Sturdy Construction Materials and Proper Reinforcement

Let's delve into the best ways to improve your coop's overall structure. It does not matter whether you have chosen robust wooden planks, resilient pallet wood, or durable compressed wood, the critical factor here is the finer details. Those spaces or seams between the boards are what truly matter in maintaining security. Over time, these seams can become vulnerable due to the weathering of the caulking used to seal them, potentially allowing probing and even full-scale attacks by predators.

To make your coop more secure, consider reinforcing the interior with crossbeams. Also, take a close look at every plank, inspecting them meticulously for knots. If you find any boards with knots, it is best to discard them promptly since these can serve as potential weak points that might allow unwanted intruders in.

The same principles of reinforcement apply to the coop's flooring. Think about adding an extra layer of protection, such as a single continuous sheet of linoleum (not individual peel-and-stick squares). Keep an eye out for knots and use crossbeams to strengthen the floor further. If your coop has a natural floor, be on the lookout for signs of digging around the outside. Raccoons, known for their digging skills, can exploit even small openings to access eggs, chicks, juveniles, and mature hens.

To counter this, consider using materials like concrete bastions, patio slabs, or retaining wall bricks as robust barriers against digging. Create a trench about one foot deep around your coop's perimeter, insert these blocks, and then cover them with soil and sod. While a determined predator might still attempt to tunnel in from a distance, these measures make it much harder and increase the likelihood that you will detect their intrusion before they get inside.

Entry points include the door to the area where you have placed the nesting boxes, the door or gate that humans use to access the coop, or the birds' pop door, which are typically the weakest parts of the coop because they are designed for easy access. To deter wild carnivores effectively, use solid wood for all three entrances.

Avoid wood paneling and plywood, as they lack the structural strength needed to withstand predator attempts. These materials also tend to deteriorate more quickly and attract lichen and mold. Ensure that the wood you use is cut to the exact dimensions of your coop to eliminate any potential weak spots.

Secure Latches and Locking Mechanisms

Secure the doors firmly with heavy-duty metal hinges and use latches with carabiners to keep them securely closed when not in use. While raccoons might be able to manipulate snap-hook locks, carabiners are typically too complex for them to handle (Hotaling, 2019).

To thwart egg hunters such as snakes, lizards, and rats, it is imperative to deny them access to your hens' nesting boxes. Ensuring secure nesting boxes means that potential intruders will leave empty-handed. To achieve this, make certain that your coop boasts lockable or well-secured nesting boxes, accessible only by human hands. While this method proves effective against most threats, it may not deter light-fingered neighbors (Smith, 2020e).

Predator-Resistant Fencing and Barriers

Predator-resistant fencing and barriers are essential for safeguarding your chicken coop against potential threats. These protective structures serve as a robust line of defense, deterring unwanted intruders. By selecting durable materials and positioning these barriers thoughtfully, you establish a nearly impenetrable shield that provides continuous protection for your feathered companions. Whether you are dealing with cunning foxes, elusive raccoons, or other troublesome creatures, a well-constructed fence or barrier system ensures your chickens can freely roam without safety concerns.

Electric fencing proves particularly effective for those facing active threats like foxes, dogs, and feral cats. Setting up this fencing is a straightforward process, and its advantages extend beyond mere security. It allows your hens to forage safely, thus reducing the expenses associated with feed and flock replacement.

Additionally, it acts as an extra layer of defense during the night, deterring nocturnal threats such as rodents and feral cats. In essence, electric poultry fencing serves as the ultimate protection for your feathered companions, ensuring their well-being and affording you peace of mind.

Nighttime Protection

Address the specific challenges posed by nocturnal predators and protect your chickens during the night.

Coop Lighting and Motion Sensors

Illuminating the area around your coop can be a powerful deterrent for nocturnal predators. Install well-placed outdoor lights, preferably motion-activated ones, around the coop. When these lights detect movement, they will flood the area with light, potentially startling and scaring away any lurking threats. Make sure the lights are positioned to cover the entire coop and its surroundings effectively.

Adding motion sensors to your coop lighting system is an excellent way to enhance security. These sensors can detect movement and trigger the lights or alarms. When a predator approaches, the sudden illumination can disrupt their activities and discourage them from attempting to breach your coop.

Secure Night-Time Confinement Methods

One of the most reliable methods to protect your chickens at night is to secure them inside a well-built coop. Invest in a coop with sturdy doors and latches. Ensure that these doors are predator-proof and cannot be easily manipulated. Additionally, consider automatic coop doors that close at dusk and open at dawn, providing an extra layer of protection. These doors can be set to a timer or linked to light sensors, ensuring your chickens are safe when darkness falls.

If you allow your chickens to roam free during the day, installing nighttime fencing will be essential. Electric poultry netting or mobile fencing can be useful for controlling your chickens' movements during the night. Ensure the fencing is predator-resistant and extends below ground to prevent digging. This way, even if a predator attempts to dig under the fence, they will be met with a barrier that discourages further intrusion.

Noise Deterrents and Scare Tactics

Embrace modern technology to bolster your protective measures. You could incorporate innovative devices like automatic coop doors, equipped with predator motion detection capabilities, and the added function of dispatching email alerts. Other options to enhance your coop's security infrastructure are solar lights and wildlife cameras. Solar lights are designed as deterrents for nocturnal prowlers, while wildlife cameras are ideal for remote surveillance.

Some nocturnal predators are sensitive to loud noises, so you could install noise deterrents, such as motion-activated alarms or even a radio set to a talk station, near your coop. When a predator approaches, the sudden noise can startle them and make them think twice about approaching your chickens. Be mindful of your neighbors when using noise deterrents and set them to activate only during nighttime hours.

Visual deterrents like reflective objects, scarecrows, or even motion-activated lights with startling patterns are also effective deterrents. Place these around your coop to create an environment that is less inviting to predators. While these methods may not work on all predators, they can be a valuable part of your overall nighttime protection strategy.

By implementing these measures, you can significantly reduce the risk that nocturnal predators pose and provide a secure nighttime haven for your chickens. These strategies not only protect your flock but also promote your peace of mind as a conscientious chicken keeper.

Natural Deterrents

In your quest to safeguard your cherished flock, nature can be a valuable ally. Natural deterrents offer a harmonious approach to predator prevention, creating an environment that discourages potential predators from encroaching upon your chicken habitat. By integrating these methods, you not only enhance the safety of your feathered companions but also foster a balance that respects the ecosystem.

Predatory Animal Deterrents, Such as Dogs or Geese

Certain dog breeds, such as Maremma or Great Pyrenees dogs, are renowned for their effective protection of hens in a coop. However, even a family dog can serve as a reliable poultry guardian for your chicken coop. Additionally, you could consider using decoy animals, such as fake owls, which are great deterrents for hawks and eagles.

Bird scare tape consists of holographic material that is intentionally designed to disrupt the navigation of avian predators by creating an unpredictable light pattern. You can achieve a similar effect by hanging old CDs around your chicken enclosure. Other effective methods to deter avian threats include using pinwheels, disco balls, or installing anti-hawk netting. Anything flashy, reflective, or shiny should help.

Safe Free-Ranging Practices

Offering adequate cover for free-ranging chickens is a crucial safety consideration. Chickens, while foraging freely, need places to seek refuge from both the elements and potential predators. This cover can include natural elements like dense bushes, shrubs, or trees that provide shade, protection from rain, and places to hide. Additionally, man-made structures like well-placed coops or chicken tractors can serve as secure shelters.

These shelters not only offer a sense of security for your chickens but also aid in minimizing stress, ensuring they can roam, scratch, and peck in a safe and comfortable environment.

Supervised Free-Ranging During Daylight Hours

Supervised free-ranging during daylight hours is an excellent way to strike a balance between allowing your chickens the freedom to explore and ensuring their safety. By keeping a watchful eye on your flock as they roam, you can quickly intervene in case there are any potential threats from predators. This approach allows your chickens to engage in natural behaviors like scratching for insects and taking dust baths while enjoying the fresh air and sunshine. Create a designated free-ranging area within your yard or garden where you can supervise their activities, ensuring they remain within a secure and protected environment.

Providing Secure Hiding Places and Escape Routes

Creating secure hiding places and escape routes is a proactive measure to safeguard your free-ranging chickens. Natural cover such as dense bushes, shrubs, or trees can offer your chickens shelter from aerial predators and a quick escape route if needed. Additionally, strategically placed man-made structures like coop outbuildings or chicken tractors can serve as safe retreats. These hiding spots not only provide protection but also reduce stress levels among your chickens, allowing them to explore with confidence, knowing they have a safe haven nearby.

Utilizing Mobile Fencing or Electric Netting

To facilitate controlled free-ranging, consider using mobile fencing or electric netting systems. These portable enclosures provide a secure perimeter for your chickens to roam while also deterring predators. You can easily move mobile fencing to fresh foraging areas, which helps to prevent overgrazing and keeps your flock safe from lurking threats. Electric netting delivers a mild shock to would-be predators, dissuading them from attempting to breach the enclosure. When used in conjunction with supervised free-ranging, these solutions offer an ideal balance between security and freedom, ensuring your chickens enjoy the benefits of outdoor exploration while staying protected.

Managing Predator Attacks and Their Aftermath

Encountering a predator attack on your cherished flock can be a harrowing experience. However, if you are prepared and have a clear plan of action you can significantly mitigate the damage and ensure the safety of both you and your chickens. This section outlines the essential steps in managing such crises effectively, safeguarding your feathered companions, and upholding the integrity of your coop.

Responding to Predator Attacks

Predator attacks can be harrowing and distressing affairs. Knowing your flock has been endangered or harmed is a keeper's worst nightmare. When an attack or coop invasion does occur remember to always prioritize your personal safety at all times before responding. Chickens may have gotten hurt but you don't want to add yourself to the casualty list. Attacks are not typically frequent and many chicken keeps will go their lifetime without ever experiencing one. However, if a predator does get to your flock, assessing your coop can tell you how the predator entered and will give you insights into preventing the same event from occurring twice.

Assessing the Situation and Ensuring Personal Safety

Your safety takes precedence above everything else. When confronted with a predator attack, take a moment to assess the situation in detail. Ensure you are at a safe distance from any immediate danger before proceeding. Remember that a calm and composed approach is crucial at this juncture. Panicking can lead to hasty decisions that may not be in the best interest of your chickens or yourself. Make use of any safety equipment or tools you have on hand, such as a flashlight, a stick, or a loud noise-making device, to deter the predator.

Chasing Away Predators Without Endangering Yourself or the Flock

It is natural to want to protect your chickens by chasing away the intruder. However, doing so without a proper strategy can pose risks both to you and your feathered companions. Predators can be unpredictable and may become aggressive when cornered. It is crucial to learn the most effective and safe ways to deter predators. If you have a dog trained for this purpose, allow it to do its job. If not, consider acquiring one. Remember that some predators, like birds of prey, are protected by law, so consult your local wildlife regulations before taking any action.

Administering Immediate Medical Aid to Injured Chickens

If your chickens fall victim to a predator, it is imperative to act swiftly, especially when injuries are involved. Predators can cause various injuries, from superficial scratches to more severe wounds. Administering immediate medical aid can mean the difference between life and death for your chickens.

Assessing and Reinforcing Coop Security

Following a predator attack, you will need to do a meticulous evaluation of your coop's security. This process allows you to identify and promptly resolve vulnerabilities and substantially mitigates the risk of future attacks. You can start by gauging the overall effectiveness of your existing security protocols, particularly in light of the recent incident. The next step involves identifying weak points and vulnerable areas where predators successfully breached your coop's defenses.

With weak points duly identified, the focus naturally shifts toward bolstering your coop's structure and fencing. Here, the key objective is to use sturdy materials that serve as robust deterrents to potential predators. You may also need to invest in upgraded locks and latches, designed to resist tampering. The overarching goal centers on crafting a coop akin to a fortress, one where predators encounter virtually insurmountable barriers.

Although they are potentially fatal to your flock, each predator attack presents a valuable opportunity to elevate your coop's security. Through a comprehensive assessment and reinforcement process, you can curtail the likelihood of future predator threats, ultimately safeguarding the well-being of your beloved flock.

Dealing With Lost Chickens

Take a gentle approach when dealing with lost chickens and all the emotions that come with it. Losing our feathered friends to predator attacks can be emotionally challenging. There are practical matters to deal with too, such as handling the remains and rebuilding the flock. In this section, we are here to lend a friendly hand and offer guidance on navigating the emotional side of such an experience while smoothly handling the practical aspects that follow.

Coping With Loss and Emotional Impact

Losing a chicken can be a heart-wrenching experience for any poultry keeper. These birds often become more than just livestock; they are beloved members of our flocks, each with its own personality and charm. When a predator strikes and we find ourselves facing the aftermath, it is entirely normal to feel a range of emotions—sadness, frustration, and even anger.

It is essential to acknowledge and process these feelings. Grieving the loss of a chicken is a valid response, and there is no set timeline for how long it should take. Everyone copes differently. Some may find solace in sharing their experiences with fellow chicken keepers, while others might need some quiet time to reflect. What matters most is allowing yourself the space to mourn in a way that feels right for you.

Proper Disposal of Chicken Carcasses

Once the initial shock and sadness start to subside, practical matters will require your attention. This includes addressing what to do with the remains of the chicken, which should be handled with care as per local regulations. Proper disposal is not only respectful but also essential for preventing further attraction of predators.

Rebuilding the Flock and Sourcing New Chickens

Rebuilding the flock is a process that many chicken keepers go through after a predator attack. It can be an optimistic step, but it also requires careful consideration. You will need to source new chickens from reputable suppliers, ensuring they are healthy and free from any potential diseases

Maintaining Vigilance and Monitoring

Maintaining vigilance will always be your first line of defense against cunning predators. Regular checks and inspections become your trusted allies, helping you safeguard your feathered friends. But remember, you are not alone in this. Building a strong and supportive community is just as essential. Here, we explore the art of vigilance and collaboration in protecting your cherished flock.

- Regular checks and inspections: Establish a routine for monitoring your chicken habitat. Regularly inspect your coop's integrity and security measures to ensure they are predator-proof.

- Scanning the perimeter: Take a walk around the perimeter of your coop and outdoor space, scanning for any signs of predator activity.

- Monitoring chicken behavior: Keep a keen eye on your chickens. Changes in behavior and stress indicators can be early warning signs of predator presence.

- Community and neighborhood involvement: Forge connections with fellow chicken keepers in your neighborhood. Sharing information and experiences is a valuable resource in staying vigilant.

- Establish neighborhood watch systems: Consider collaborating with your neighbors to establish neighborhood watch systems. Together, you can keep an eye out for potential threats.

- Collaborating on deterrent strategies: Work together with your community to develop and implement effective predator deterrent strategies. Collective efforts can significantly reduce the risk to your chickens.

Interactive Exercise

Conduct a predator risk assessment for your chicken habitat. Identify potential vulnerabilities, such as gaps in fencing, unsecured coop access points, or attractants like food scraps. Develop a plan to address these vulnerabilities and implement preventive measures to minimize the risk of predator attacks.

Chapter 11:

Managing Chicken Waste

In this chapter, we embark on a journey to master the art of waste management in your chicken haven. We will explore effective methods to keep your chicken habitat clean and odor-free while also minimizing your environmental footprint. Get ready to discover practical solutions for handling waste with finesse and learn how to harness the potential of chicken manure as a valuable resource for your homestead.

Importance of Waste Management in Chicken Raising

Effective waste management is crucial, especially in smaller backyards. Fresh chicken manure poses health risks due to the harmful bacteria, high ammonia content, and odors it releases. However, chicken waste can be converted into a valuable resource through composting and proper composting will help to eliminate these risks and provide exceptional benefits to your garden. Failing to compost can contaminate soil and harm plants while inadequate waste management can harm chickens, leading to ammonia-related health issues and discomfort. Composting reintroduces organic matter and nutrients into the soil, enhancing its vitality. Methods like the deep litter system and organic bedding materials contribute to a sustainable partnership between chickens and gardens, embodying permaculture principles.

Understanding the Impact of Chicken Waste

For those taking care of a standard backyard ensemble of six chickens, the accumulation of manure can indeed become quite substantial over a year. While this situation may be readily managed on expansive homesteads, it does present a unique challenge in the confines of urban and suburban backyards. The solution lies in astute management, particularly in converting this seemingly burdensome waste into a resource as valuable as the eggs your hens provide. With a bit of extra effort and a good grasp of the composting process, you can indeed master the art of converting chicken manure into a rich garden asset.

Many chicken keepers are already aware of the potential health hazards associated with fresh chicken manure. These concerns include the presence of harmful bacteria like Salmonella or E. coli, the high ammonia content which is impractical for direct use as a fertilizer, and the off-putting odor that lingers around fresh manure. However, the revelation lies in the transformation that properly composted chicken manure undergoes. Through the composting process, the offensive odors are eliminated, rendering it not only safe but also exceptionally beneficial (Garman, 2021).

Environmental Pollution and Contamination Risks

Introducing manure directly into your garden, without following proper composting procedures, presents a potential risk. It can lead to pathogenic organisms being disseminated within the soil and then absorbed by low-lying leafy greens and fruit-bearing plants. It is vital to understand that fresh manure is notably potent and can inflict harm on both plant roots and leaves due to its elevated nutrient content. Therefore, you always need to follow the requisite composting procedure to mitigate these risks and ensure the safety of your garden (Garman, 2021).

Negative Effects on Chicken Health and Well-Being

Neglecting to compost chicken manure also increases ammonia levels, promotes pathogen growth, attracts pests, and creates odors. This harms chicken health and can result in respiratory issues, diseases, and discomfort for the chickens. It also impacts environmental health, leading to pollution through runoff (Garman, 2021). Proper composting is essential for maintaining chicken well-being and preventing these problems.

Utilizing Waste as a Valuable Resource For Your Garden

Compost derived from chicken manure can be a potent amendment to your garden soil. It effectively reintroduces essential organic matter into the soil matrix while significantly enhancing its nutrient profile. Key elements such as nitrogen, phosphorus, and potassium are released in a balanced and slow-release manner, making it a highly prized addition to any garden (Garman, 2021). In this way, the management of chicken waste evolves from a challenge to a strategic resource for nurturing and enhancing your garden's vitality and productivity.

Different Methods of Chicken Waste Management

Chicken waste management encompasses three primary approaches: composting, deep litter, and vermicomposting with chicken manure. Composting involves the decomposition of chicken manure, bedding materials, and other organic waste into nutrient-rich compost. This process enhances soil fertility and reduces waste. Deep litter systems entail creating a bedding layer within the coop, which gradually breaks down with the help of chickens, fostering a rich humus that improves the environment and offers potential for composting. Vermicomposting, on the other hand, employs specially selected earthworms to digest chicken manure and organic matter, ultimately producing valuable worm castings rich in nutrients. Each method offers unique advantages in managing chicken waste while contributing to sustainable and eco-friendly practices in poultry keeping.

Deep Litter Method

Introducing chickens to your permaculture garden presents a mutually beneficial partnership. They excel at pest control, consuming common garden nuisances, and gaining essential nutrients. Additionally, their organic manure-enriched compost can be brewed into nutrient-rich manure tea which can be added to garden soil. Spent coop bedding, especially organic options like hemp bedding, enhances soil quality when added to compost or used as mulch.

Consider the deep litter method with organic hemp bedding if you want to create a self-contained composting system within the chicken run, further reduce waste, and produce nutrient-rich material for your garden. This collaboration establishes a sustainable cycle: Chickens contribute pest control and waste products, while your garden provides essential resources and sustenance for your birds. It embodies the core principles and mutual benefits of permaculture.

Layering and Maintaining the Litter Bed

When venturing into deep litter composting, the best time to start the process is in early autumn, as it allows the bedding to start decomposing before the onset of a harsh winter. Begin by removing any existing bedding, particularly if sand or non-absorbent straw is present. For concrete floors, apply a 1-inch layer of potting or garden compost to establish a suitable earth base for absorption and microbe activity. The initial layer of bedding should be carbon-based, absorbent, and fine enough to facilitate rapid decomposition. Materials like wood shavings, untreated and unscented, are recommended. Aim for a depth of 4 to 6 inches, preventing bedding from scattering by placing a 6-inch-high wooden plank near coop entrances (Andrews, n.d.). To monitor depth conveniently, this plank can serve as a visual indicator.

Balancing Carbon-Rich and Nitrogen-Rich Materials

After the initial layer, regularly add 1 inch of fresh bedding every few days without removing the existing layer. Adjust the frequency according to moisture levels; if it becomes excessively damp or exhibits a nitrogen odor, add more dry bedding. Make sure to use materials that are free from pesticides or chemicals, such as dried leaves, chopped straw (in moderation due to its low absorbency), grass clippings, pine needles, chipped bark, or bamboo. Bamboo, if available, can serve as an excellent sustainable source for bedding. Be careful when using diatomaceous earth, as its use can disrupt the microorganisms vital for the composting process (Andrews, n.d.).

Managing Moisture Levels and Odor Control

Maintaining proper moisture balance is challenging yet pivotal in deep litter systems. While some moisture is necessary for composting, excessive dampness can lead to respiratory and health issues. To control moisture, position waterers outside the coop to prevent your chickens from drinking at night. Ensure adequate ventilation in your coop to allow for proper drying but make sure not to create any drafts. Regularly break up any clumps in the bedding and integrate them into the rest. To assess moisture levels, squeeze a handful of bedding; it should hold its shape and crumble into a rich humus without dripping water. Lastly, keep your olfactory senses sharp; any ammonia smell signifies inefficient composting and requires cleaning (Andrews, n.d.).

Composting Chicken Manure

Efficiently managing a chicken coop involves more than just keeping your feathered friends happy and healthy; it also entails effective waste management. Learning how to clean a chicken coop and what to do with the waste is a crucial aspect of responsible poultry keeping. A vital component of waste management is composting chicken waste. From the materials needed to the recommended compost ratios and essential temperature checks, we will unravel the secrets of transforming chicken coop waste into a valuable resource for your garden while ensuring safety and sustainability in the process.

Proper Composting Techniques and Best Practices

Any chicken owner seeking to make the most of their coop waste needs to master proper composting techniques and know how to adhere to best practices. Composting effectively hinges on achieving the right balance of components, typically categorized as brown and green materials. In this context, bedding materials like shavings, sawdust, straw, and hay, combined with yard plant debris, leaves, small sticks, and paper, constitute the brown components.

On the other hand, the green elements encompass fresh manure and kitchen scraps. The first step is to achieve an optimal compost ratio, typically recommended at two parts brown to one part green due to the manure's high nitrogen content. Once you have gathered these materials, place them in a suitable compost bin or composter, for which the recommended size is a cubic yard (Garman, 2021).

Compost Pile Management and Maintenance

To encourage decomposition, you will need to regularly mix, stir, and turn the compost pile to ensure that air reaches the core helping beneficial bacteria to break down the pathogens in the compost. You will also need to monitor the inner core temperature, making sure it stays between 130°F and 150°F, in order to neutralize harmful bacteria. With patience and diligent management, it should take approximately a year to yield nutrient-rich compost, primed for enhancing your garden's vitality. While the heat generated during composting should eliminate pathogens like E. coli and Salmonella, it remains advisable to exercise caution and thoroughly wash any produce cultivated in a compost-fed garden (Garman, 2021).

Utilizing Chicken Manure Compost in Your Garden

Chicken manure compost revitalizes your soil by infusing it with crucial nutrients such as nitrogen, phosphorus, and potassium. Your plants will thrive, fostering robust root systems and yielding abundant harvests. In this sustainable cycle, your chickens contribute to your garden's prosperity, while your garden enhances your chickens' living conditions.

Vermicomposting

Vermicomposting, or vermicompost, is an organic technique that employs earthworms, mainly red wigglers like Eisenia fetida, to convert organic waste into nutrient-rich compost. In specialized containers like worm bins, these earthworms consume materials such as kitchen scraps, yard waste, and paper, digesting them into dark, crumbly worm castings. This vermicompost is a potent soil conditioner and fertilizer, brimming with nutrients and beneficial microorganisms. It enhances soil structure, moisture retention, and plant growth, making it ideal for eco-conscious gardeners and indoor composting. Vermicomposting also reduces landfill waste and can provide high-protein worm feed for poultry, offering an efficient and sustainable organic waste management solution (Pickett, 2013).

Setting Up a Vermicomposting Bin

Setting up a vermicomposting bin is a relatively straightforward process. Start by selecting an appropriate container, which can be a plastic or wooden box or a specialized worm bin designed for this purpose. Make sure it comes with a lid to keep pests out. Next, prepare your bedding material, which can consist of shredded newspaper, cardboard, coconut coir, or a mixture of these. Moisten the bedding until it reaches the consistency of a wrung-out sponge.

With your container and bedding ready, it is time to introduce the stars of vermicomposting: red wiggler worms, scientifically known as Eisenia fetida. These worms are excellent composters, consuming substantial amounts of organic matter and reproducing rapidly.

To set up the bin, drill small holes in the container's bottom and sides to provide aeration and drainage, placing a tray or container beneath to collect any liquid runoff. Fill the container with the moist bedding material, fluffing it up to ensure proper airflow. Then, add your red wiggler worms on top of the bedding, starting with approximately one pound (roughly 1,000 worms) for every square foot of surface area.

The bedding must remain consistently damp but not soaked or waterlogged and should be kept at a stable temperature within the bin, ideally between 55°F and 77°F (13°C to 25°C). This moisture level and temperature range ensure the worms stay active and productive (Pickett, 2013).

Over time, the bedding will turn into nutrient-rich worm castings. The method to harvest these castings is easy: simply move the worms to one side of the bin. Then, add fresh bedding and food to the opposite side. The worms will naturally migrate to the new bedding, leaving behind the castings.

Feeding and Caring for Composting Worms

Once your vermicomposting bin is established, maintaining it is relatively straightforward. Feed the worms regularly with small amounts of organic waste like kitchen scraps, coffee grounds, and crushed eggshells. Bury the food scraps beneath the bedding to prevent pests and odors. Remember that worms can consume around half their body weight in food each day, so avoid overfeeding them (Pickett, 2013).

Controlling Odor in Your Chicken Habitat

As loveable as they are, chickens are stinky. Unfortunately, there is no quick and easy hack to eliminate chicken coop odor for good. However, with regular maintenance, dedication, and the right amount of ventilation, your coop should not cause too much offense to the olfactory senses.

The only surefire way to take care of a stinky coop is to keep it clean, but if you are still bothered by your flock's natural bouquet, then there are a few things you can do to help: ensure your coop has proper ventilation, keep your coop dry, clean the coop regularly, and place fresh herbs in bedding and plant them near the coop.

Proper Ventilation

In Chapter 3, we discussed coop ventilation and ventilation options, we also went over the importance of airflow for chicken respiratory health and biosecurity in Chapter 6. Here we will provide a brief overview of that content as it pertains to waste management.

Ventilation Requirements for the Coop

Effective coop ventilation is a non-negotiable element for maintaining a healthy and comfortable environment for your flock. The primary purpose of coop ventilation is to ensure the circulation of fresh air while preventing the buildup of harmful ammonia gasses and unpleasant odors. It is essential to strike a balance, providing enough ventilation to promote air exchange without creating drafts that can lead to temperature fluctuations and stress among your chickens.

Natural and Mechanical Ventilation

There are various ways to achieve proper coop ventilation, with both natural and mechanical options at your disposal. Natural ventilation methods include pop holes and large openings covered with mesh, ideal for smaller coops. However, they may lack closing mechanisms that are essential during colder months, requiring additional measures like using sheets or curtains to retain warmth. Another natural option involves creating ventilation holes in the coop ceiling, preferably on the north and south ends, and covering them with mesh. This facilitates the flow of fresh air while keeping unwanted pests out.

For those seeking a more automated solution, mechanical options like turbine ventilation systems can be installed on the coop's roof. These systems effectively remove trapped gasses and excess moisture. Still, remember that even with mechanical ventilation, cross-ventilation, usually achieved through windows, remains crucial. Multiple windows in your coop allow you to adjust airflow according to the season, invite sunlight, and provide essential natural light to your chickens.

Options Monitoring and Adjusting Ventilation Levels

Maintaining the right ventilation levels in your coop is an ongoing process. You will need to monitor it regularly to ensure that your chickens are getting the fresh air they need while avoiding drafts and excessive temperature fluctuations. Signs of respiratory distress or unusual odors can serve as indicators of improper ventilation.

To fine-tune your coop's ventilation, consider making adjustments based on the season and weather conditions. For instance, during hot summer months, you might want to increase airflow, possibly by using higher ceilings to improve circulation. In colder winters, you can reduce ventilation to maintain warmth while avoiding drafts. Balancing coop ventilation effectively contributes to odor and humidity control (Smith, 2020b).

Keep a moisture meter in your coop to check the humidity levels. If the readings are too high then you know more ventilation is needed. Another option is to take a whiff, if the air smells or feels stale then it is time to open a window or two.

Regular Cleaning and Maintenance

Maintaining a clean coop will keep both you and your chickens happy. When you establish a regular cleaning routine and adopt effective odor-reduction strategies you can significantly improve the air quality in your coop.

Removing Soiled Bedding and Waste

Effective waste management, including regularly removing soiled bedding and waste, is at the core of odor reduction in your coop. Identify areas where bedding has become soiled and where waste has accumulated and use a rake, shovel, or scoop to carefully remove these materials. You then need to make sure that you follow proper disposal practices, whether through composting or as per local waste regulations to avoid any contamination or negative environmental impact (Chicken Whisperer Magazine, n.d.).

Cleaning Coop Surfaces and Nesting Areas

You will also need to regularly clean the coop surfaces and nesting areas for effective odor control. Remove all bedding and debris from these surfaces. Use a suitable cleaning solution or a mixture of water and vinegar to disinfect. Pay special attention to corners, crevices, and nesting boxes. Thoroughly scrub and rinse all surfaces, allowing them to air dry before adding fresh bedding.

Sanitizing Feeders and Waterers

Clean feeders and waterers help to prevent odors associated with mold, algae, and contamination. Regularly inspect these containers and disassemble them for thorough cleaning. Use a mild detergent and a brush to remove any residue and then rinse and sanitize them well with a poultry-safe disinfectant. Clean containers also help maintain food and water quality in addition to reducing odors in the coop.

Natural Odor Control Methods

Explore natural methods to combat and neutralize odors in and around your chicken habitat. These include options such as using easy-to-clean materials, such as sand, wood shavings, or straw. Or, incorporating chicken plants and herbs, such as rose petals, lavender, lemon balm, or sage.

Do not use deodorizers in your coop as these can be harmful to your chickens, even natural deodorizers such as baking soda can be hazardous to keep in the coop with your feathered friends (Chicken Whisperer Magazine, n.d.).

Safe Handling and Disposal of Chicken Waste

Proper Waste Disposal

It is important to learn the responsible methods of handling and disposing of chicken waste to prevent contamination and environmental harm. By following these guidelines for safe handling and storage, composting, and creating chicken manure tea, you can utilize chicken manure as a valuable and eco-friendly fertilizer for your garden or crops while minimizing potential risks. No matter which waste disposal strategy you choose, always adhere to local regulations regarding manure management.

Following Local Regulations and Guidelines

Adhering to local poultry waste guidelines and regulations not only ensures legal compliance but also contributes to responsible environmental stewardship. Proper waste management benefits your operation, neighboring communities, and the environment as a whole. Here are some tips to help you navigate legal compliance.

- Research local regulations: Start by researching the poultry waste regulations applicable in your state or locality. These regulations are typically available on government websites, agricultural extension offices, or environmental agencies.

- Think about the environment: Consider what surrounds your place when you are dealing with poultry waste. If you are close to water or other sensitive spots, there might be extra rules to keep the environment safe (Kyakuwaire et al., 2019).

- Contact local authorities: Reach out to local authorities or agencies responsible for poultry waste management. They can provide up-to-date information, guidelines, and resources to help you comply with regulations.

- Obtain necessary permits: Depending on the scale of your poultry operation, you may need permits or licenses to handle and dispose of poultry waste. Contact the relevant authorities to determine the permit requirements and initiate the application process if necessary.

- Comply with zoning regulations: Ensure that your poultry facility complies with zoning regulations, which may dictate the distance between your operation and neighboring properties, water bodies, or sensitive areas.

- Develop a waste management plan: Create a poultry waste management plan that aligns with local regulations. This plan should include details on waste storage, handling, disposal, and any required record-keeping.

- Stay informed and update your practices: Regulations and guidelines can change over time. Stay informed on any updates or revisions to local poultry waste regulations and adjust your practices accordingly.

Avoiding Improper Disposal Practices

It is crucial to be mindful of the environment when handling poultry waste. A golden rule is to never toss chicken waste into bodies of water or drainage systems that are close to water or that lead to bodies of water.

Whether you are disposing of chicken litter or diseased birds that have passed away, you will need to avoid contaminating the environment by all means possible. Poultry can harbor various pathogens, including bacteria, fungi, helminths, parasitic protozoa, and viruses (Kyakuwaire et al., 2019). If your chickens are not organically raised, there is an additional risk that your poultry waste may contain growth hormones, heavy metals, pesticides, and antibiotics which can perpetuate antibiotic-resistant pathogens.

Many of these health risks can spread through water contamination and lead to an epidemic—or worse, a pandemic. Because of the risk these elements can pose to humans and the environment, you should always be prudent and responsible when handling your flock's waste.

Utilizing Waste Management Services (If Available)

If you do not want to compost your poultry waste, have a look in your area for companies that collect manure, or local farmers, homesteaders, and gardeners who are looking to add nutrients to their soil and are willing to pick up your chicken poop for free. Just make sure that, in case you do have to store it while waiting for collection, that you do so properly—you can read more about storage options below.

Utilizing Chicken Manure as Fertilizer

Safe Handling and Storage of Chicken Manure

It is important to follow best practices for the safe handling and storage of chicken manure to harness its benefits while making sure to minimize environmental risks.

When handling chicken manure, always wear appropriate protective gear such as gloves and a mask to avoid any direct contact or inhaling any dust or pathogens.

After cleaning the coop and collecting manure, transfer it to a designated storage area using a shovel or rake. Avoid spreading it directly on crops or garden beds if it has not been processed properly. Always wash your hands thoroughly after handling chicken manure to reduce the risk of contamination.

A few storage options for your chicken manure are (Cunningham et al., 2003):

- Compost: One of the safest ways to handle chicken manure is by composting it. Composting not only reduces pathogens but also transforms the manure into a valuable soil conditioner.

- Covered stockpiles: If you need to store fresh manure temporarily, consider covered stockpiles. Cover the piles with plastic sheeting anchored securely to prevent rainwater infiltration and minimize odor. Make sure your stockpiles have ground liners so that there is no risk of the uncured manure leaching into the soil.

- Permanent storage structures: For larger operations, permanent storage structures with concrete floors and roofs provide a safe long-term solution. These structures protect the manure from the elements, reducing the risk of leaching and overheating.

Compost Application Guidelines

Allow the compost to cure for several months to ensure that it is fully stabilized. Properly cured compost is rich in nutrients and free of pathogens and should be ready for direct application to your garden beds. Always keep your compost, whether you have a heap or a bin, in a dry area free of rain.

Incorporating Chicken Manure Tea or Liquid Fertilizer

You can create chicken manure tea by steeping well-aged, composted manure in water. This liquid fertilizer is rich in nutrients and can be applied directly to plants.

Use a ratio of one part composted chicken manure to five parts water. Place the manure in a cloth bag or mesh container, submerge it in the water, and allow it to steep for several days to weeks, while stirring it occasionally.

Dilute the chicken manure tea with water at a ratio of 1:10 (tea to water) before applying it to your plants or garden (Cunningham et al., 2003). This prevents over-fertilization, which can harm plants, while still providing a good amount of nutrients to your plants.

Apply the diluted tea to the base of plants or as a foliar spray, preferably during the growing season to provide a nutrient boost.

Monitoring and Troubleshooting Waste Management

Monitoring Waste Accumulation

Establishing a robust monitoring system will help ensure that you are following responsible waste management in your coop. This entails conducting regular waste assessments and observations to monitor accumulation, identify issues, and make necessary adjustments.

Regular Waste Assessments and Observations

Regular waste assessments involve physical inspections of your coop and its surroundings. Look for visible indicators of waste buildup, such as soiled bedding, excessive droppings, or areas with unpleasant odors. Evaluate your coop's condition comprehensively, and inspect the quality of the litter, moisture levels, and the overall health and behavior of your chickens. By conducting these assessments at consistent intervals, you can promptly detect and address potential problems.

Identifying Signs of Excessive Waste or Improper Management

Recognizing signs of excessive waste accumulation or improper waste management will help you address any issues early on. These signs may manifest as unpleasant odors, elevated moisture levels, deteriorating chicken health, infestations of insects and pests, as well as the growth of mold and mildew. Once you have identified such signs, take swift action to rectify the situation. This might involve increasing the frequency of cleaning, enhancing ventilation, ensuring proper waste disposal, adjusting bedding materials, and optimizing feeding practices (Amit, 2020).

Making Necessary Adjustments to Maintain Cleanliness and Odor Control

Monitoring waste accumulation is a fundamental component of responsible poultry management. Through regular assessments and timely adjustments, you can maintain a clean, healthy, and odor-free environment for your chickens. This not only promotes their well-being but also contributes to their productivity and overall quality of life.

Troubleshooting Common Waste Management Challenges

Managing chicken manure is not without its challenges. To maintain an effective waste management system, it is crucial to address these common issues with practical solutions. Some common issues are excess moisture, ammonia build-up, pest infestations, and maintaining nutrient balance.

By addressing these common challenges with practical solutions, you can maintain a healthy composting system.

Excessive Moisture or Ammonia Levels

Excessive moisture in your compost pile can hinder the decomposition process and lead to unpleasant ammonia odors. To combat this issue, consider turning the compost pile more frequently to aerate it and promote drying. You can also add dry, absorbent materials like straw or sawdust to balance the moisture content and monitor the moisture levels regularly to prevent ammonia buildup.

Pest Infestations

Compost piles often attract pests like flies and rodents. To deter these, ensure your compost pile is well-covered and sealed, and add a layer of straw or leaves on top to help create a barrier. Avoid putting any food scraps in the compost that might attract pests. Regularly turning the compost pile can disrupt pest habitats and discourage infestations.

Balancing Waste Materials for Optimal Composting

Efficient composting is all about achieving the right balance of waste materials. If your compost pile seems slow to decompose, you may need to adjust the mix of materials. Ensure you have a good balance of green (nitrogen-rich) and brown (carbon-rich) materials. Green materials include chicken manure and kitchen scraps, while brown materials comprise straw, leaves, and wood shavings. Properly layering and mixing these materials can promote optimal composting.

Implementing Sustainable Practices

Sustainable agriculture practices include permaculture and regenerative farming principles. These approaches aim to create agricultural systems that fulfill human needs while also benefiting the environment and local communities, enhancing food system resilience. A crucial component of these practices is regenerative grazing, which plays a pivotal role in soil health improvement while managing livestock on perennial and annual forages (Wallace Center, n.d.).

As modern agricultural practices became more specialized and less diverse, the way we were farming started disrupting nutrient cycles. This led to higher costs for moving manure and bringing in synthetic fertilizers. Regenerative grazing, when done thoughtfully, can reverse this trend. It not only nurtures soil health and reduces nutrient loss but also reduces the need for external inputs, diversifying farm income. These practices offer solutions to enhance sustainability and resilience in agriculture and help re-establish healthy nutrient cycles on small and larger scales (Wallace Center, n.d.).

Incorporating Permaculture and Regenerative Farming Principles

Regenerative grazing is not a rigid set of rules but rather a flexible approach. It involves managing livestock intensively with frequent rotations and allowing pastures extended periods of recovery. Animals graze intensely in one area before being moved to another, giving the land and plants time to rejuvenate. Key elements include minimizing or eliminating synthetic inputs like artificial fertilizers and plowing, promoting diversity among plants, animals, and microbes, and generating sufficient revenue to support profitable farms and fair labor compensation.

Utilizing Chicken Foraging and Grazing Techniques

Chickens play an essential role in restoring closed nutrient cycles in integrated farming systems. Historically, farms integrated livestock and crops, with animals grazing and enriching the land through their manure. Chickens, with their nutrient-rich droppings, can significantly contribute to soil health and reduce reliance on external inputs like synthetic fertilizers. This integrated approach enhances farm resilience and offers environmental benefits.

Integrating Chickens Into Garden Systems

Integrating chickens into regenerative farming means raising them alongside perennial crops and trees. This approach enriches the soil with organic matter and carbon improves water retention and enhances land resilience in the face of climate challenges. Unlike mobile barn-based poultry production, this method entails keeping birds in one place, where they work harmoniously with perennial crops to improve overall land health (Wallace Center, n.d.).

Interactive Exercise

Conduct a waste management audit in your chicken habitat. Assess your current waste management practices, identify areas for improvement, and develop an action plan to implement effective waste management techniques. Consider factors such as waste accumulation, odor control, and how you could use waste as a resource.

Chapter 12:

Chicken Breeds and Varieties

In the world of chicken breeds, feathers tell stories, and clucks reveal histories. These feathered friends have long been a part of our farms, not only for their meat and egg-laying abilities but also for their companionship. In this chapter, we will take an in-depth look at the diverse landscape of chicken breeds. Much like humans, chickens come in all shapes, sizes, and temperaments. From the practical to the eye-catching ornamental, these remarkable birds offer something for everyone. Whether you are an experienced poultry keeper or just considering your first flock, learning about the various chicken breeds and their recognizable traits is a key part of chicken raising.

Exploring Popular Chicken Breeds

Profiles of Different Chicken Breeds and Their Characteristics

In Chapter 1, we discussed a few different types of chickens including some of the most common egg-layers, dual-purpose, and broiler breeds. This section will provide a more in-depth look into the world of chicken breeds as well as a brief overview of information covered previously.

Selecting Breeds Based on Climate, Purpose, and Personal Preferences

When it comes to choosing the right chicken breed for your flock, several factors should guide your decision. Climate, purpose, and personal preferences all play essential roles in selecting the perfect breed. Whether you are aiming for egg or meat production or ornamental beauty, there is a chicken breed suited to your needs.

Your local climate will also affect which breeds you choose. Chickens are sensitive to temperature and weather conditions. Some breeds thrive in hot climates, while others are better adapted to cold regions. Before bringing home a flock, assess your local climate and choose breeds that can withstand the conditions. This ensures your chickens remain healthy and productive year-round.

Your purpose for keeping chickens is another significant factor. Are you raising chickens primarily for eggs, and meat, or as charming additions to your backyard? Different breeds excel in each category. For egg production, breeds like the Rhode Island Red, Leghorn, or Sussex are popular choices. If meat is your goal, consider Cornish Cross or Broiler breeds, known for their meaty characteristics. Those seeking ornamental chickens for their unique appearances should explore breeds like the Silkie, Polish, or Serama.

Personal preferences matter as well, of course. Some people have a particular fondness for certain breeds due to their appearance, temperament, or historical significance. Take some time to research different breeds and consider the aspects that appeal most to you. Whether you are captivated by the elegance of a Legbar's blue eggs, the striking plumage of a Wyandotte, or the docile nature of a Buff Orpington, personal preferences can be a significant influence in your selection process.

Rare and Heritage Chicken Breeds

In the realm of poultry heritage, some chickens stand as living testaments to a bygone era, much like heirloom vegetables that hark back to times when supermarkets were not overflowing. These heritage chicken breeds are akin to the chickens your grandparents might have raised on their farms. They are the ancestors of modern chicken breeds and represent a vital part of poultry history.

While a few heritage breeds like Australorps and Rhode Island Reds remain fairly common, others, such as the Dutch Bantam, face endangerment even in their native Australia. Some breeds, like the Old English Pheasant Fowl, are so rare that they are at risk of extinction worldwide.

Preservation and Conservation of Endangered Chicken Breeds

Preserving the genetic diversity of heritage chicken breeds holds some significance. These heritage breeds boast a far-reaching diversity that sets them apart from the limited commercial varieties currently prevalent. This rich genetic variation is paramount for our future because it equips us to navigate potential changes or disasters.

When our chicken populations predominantly possess a narrow set of genetic traits, resembling the uniformity found in commercial breeds, their ability to adapt and thrive amidst evolving circumstances significantly diminishes. The repercussions of such a decline in chicken populations extend beyond the coop; it can potentially affect human populations as well (Rachael, 2023).

Showcasing Rare and Heritage Breeds and Their Unique Qualities

Commercial chicken breeds tend to share similar characteristics due to being bred for specific purposes. However, when it comes to heritage chicken breeds, with hundreds of unique breeds, it is tricky to make sweeping generalizations. Some of these heritage birds are prolific layers, almost rivaling their commercial counterparts, while others may not be as productive in the egg department. What is truly fascinating is that there are heritage breeds tailored to thrive in specific environments or lifestyles, whether it is handling challenging climates with ease or embracing a free-ranging lifestyle (Rachael, 2023). The world of heritage chickens is a rich tapestry of diversity.

Commercial chicken breeds exhibit remarkable uniformity and are generally optimized to thrive under specific conditions. However, when we turn our gaze to heritage breed chickens, we enter a realm of diversity and an astounding array of colors, sizes, and a wide range of comb types. Some flaunt strikingly long tail feathers, quirky top knots, or charmingly feathery feet. When it comes to egg production, they paint a vibrant canvas with eggs of varying sizes and shades, forming a veritable rainbow (Rachael, 2023).

Heritage breeds are not just diverse in appearance; they are also highly adaptable. They have evolved to flourish in diverse environmental conditions, from scorching heat to biting cold. What is more, they have been meticulously bred for various purposes, whether it is for prolific egg-laying, succulent meat production, the art of cockfighting, or simply their aesthetic appeal. This rich variety empowers you to select a chicken breed tailored to your specific needs and circumstances, ensuring that your flock thrives in a manner that aligns with your goals.

Keeping heritage breed chickens comes with a multitude of advantages:

- Wide breed selection: With such an extensive range of breeds available, there is undoubtedly a heritage chicken breed that will perfectly align with your preferences and goals.

- Extended egg-laying period: While heritage breeds may not match the commercial breeds in egg production intensity, they make up for it by laying eggs consistently over a more extended period.

- Longer lifespan: Heritage breeds tend to enjoy a longer and healthier lifespan, allowing you to enjoy their delightful company for years.

- Versatility in diet: These chickens thrive on a more diverse diet and excel at free-ranging, embracing the natural world with gusto.

- Excellent motherhood: Heritage hens exhibit exceptional mothering instincts, making them superb caregivers for their chicks.

- Environmental resilience: These birds are remarkably resilient, adapting admirably to various environmental challenges, including temperature extremes.

- Self-sustaining: If you wish to hatch your chicks, heritage breeds are your go-to choice, as they exhibit natural brooding tendencies.

- Incredible variety: The diversity among heritage breeds not only promises a kaleidoscope of egg colors but also a plethora of characteristics and traits.

In a world brimming with options, heritage breeds stand as a testament to the exquisite diversity that nature offers. Their unique qualities, both in terms of appearance and performance, make them an ideal choice for those seeking a deeper connection with their flock and the satisfaction of a self-sustaining, vibrant, and resilient chicken community.

Supporting Breed Diversity and Sustainable Farming Practices

Heritage breed chickens support breed diversity and sustainable farming by preserving genetic variety, adapting to local conditions, and promoting eco-friendly practices. Their resilience reduces our reliance on antibiotics and pesticides, and their unique qualities can create niche markets for farmers, fostering economic stability. This supports cultural traditions and maintains agricultural heritage while enhancing disease resistance and animal welfare.

Hybrid and Crossbreed Chickens

Hybrid and crossbreed chickens are both products of deliberate breeding, but they serve different roles in the world of poultry. Hybrid chickens are the offspring of carefully chosen parent breeds, often a combination of a prolific egg layer like the White Leghorn and a meaty breed such as the Rhode Island Red. They are bred with a singular commercial purpose in mind, whether it is high egg production, rapid growth, or efficient meat production.

Hybrids are recognized for their consistent size, color, and behavior, a deliberate design aimed at predictability and high productivity. Nevertheless, their productive life span is often short, especially among egg-laying hybrids, as their output tends to decrease as they age.

In contrast, crossbreed chickens emerge from the mating of chickens from different breeds, without a primary emphasis on preserving specific traits. Crossbreeding can happen intentionally or naturally in mixed flocks. Crossbreeds are commonly found in small-scale or backyard settings, and while they can be bred for certain purposes, they are inherently more versatile and adaptable to various conditions. Crossbreeds exhibit a wide range of characteristics because they inherit genes from different parental breeds, resulting in unique appearances and behaviors. They often have longer and healthier lives, especially when bred for overall hardiness and adaptability.

In essence, hybrid chickens are carefully engineered for specific traits and commercial use, while crossbreeds emerge from the blending of various breeds and tend to be more adaptable and diverse in their characteristics. Your choice between the two depends on your farming goals and preferences, as well as the scale of your poultry operation.

Understanding the Advantages and Considerations of Hybrid Chickens

Hybrid chickens have gained immense popularity in modern poultry farming for the exceptional advantages they offer. These birds are meticulously bred by harnessing the best qualities of both parent breeds. One of the primary advantages of hybrid chickens is their impressive growth rate and efficiency in converting feed into meat or eggs. They tend to reach market weight quickly, making them a preferred choice for meat production. Additionally, hybrid chickens often exhibit uniformity in size and color, which can be advantageous for commercial operations. However, it is crucial to consider that hybrids are typically bred for specific purposes, and their genetics may not be ideal for breeding future generations. This means that farmers may need to purchase new hybrid chicks regularly to maintain their performance.

Hybrid Vigor and Performance in Egg-Laying or Meat Production

Hybrid vigor, also known as heterosis, is a significant factor contributing to the remarkable performance of hybrid chickens in both egg-laying and meat production. This phenomenon occurs when two genetically diverse chicken breeds are crossed, resulting in offspring that outperform their purebred parents. In the case of egg-laying hybrids, they often exhibit impressive laying rates, producing a high quantity of eggs efficiently. For meat production, hybrid chickens grow rapidly and efficiently convert feed into muscle, resulting in excellent meat yields. However, it is important to note that hybrid vigor may not be consistently passed on to the next generation if these chickens are used for breeding purposes (Thesing, 2019). Therefore, farmers often rely on specialized breeding programs to maintain hybrid performance.

Hybrid Breeding and Management Strategies

Breeding and managing hybrid chickens require a specific set of strategies to maximize their potential. Hybrid breeding programs are highly controlled to ensure the desired genetic traits are passed onto the offspring. Typically, specialized breeding companies handle the initial crossing of the parent breeds to produce hybrid chicks.

Farmers who raise hybrid chickens for meat or egg production must carefully manage their flocks to optimize performance. This involves providing a balanced diet, maintaining ideal living conditions, and monitoring the health and well-being of the birds.

Since hybrids are often raised for commercial purposes, efficient management practices are essential to ensure a profitable poultry operation. Backyard keepers should also consider the long-term sustainability of their hybrid flocks and plan for regular replacement with newly purchased hybrid chicks to maintain high levels of productivity.

Exploring New Chicken Varieties

In the dynamic world of poultry farming, the exploration of new chicken varieties represents an exciting journey into uncharted territory. As chicken enthusiasts, farmers, and backyard keepers seek to diversify their flocks and find the perfect match for their specific needs, new chicken varieties continue to emerge. These innovative breeds, developed through careful selection and breeding, offer a fresh perspective on poultry keeping.

Whether you are in pursuit of superior egg layers, robust meat producers, or simply a breed that fits your unique preferences, the realm of new chicken varieties has much to offer. In this section, we will delve into these recent additions to the poultry landscape, shedding light on their distinctive characteristics, advantages, and the exciting possibilities they bring to the world of chicken raising.

It is worth noting that, even within the same breed, each chicken has its own special personality traits and quirks. Some breeds are versatile, serving multiple purposes because of their unique features. It is therefore important to note that the descriptions we provide are generalizations; each chicken within a breed might show these traits in its own way. Embracing this diversity, we honor the unique qualities that every feathered friend adds to the vibrant tapestry of poultry farming.

Introduction to Unique and Rare Chicken Breeds

Please note that some breeds may fall into multiple categories, as they possess unique features and serve specific purposes. Chicken personalities can also vary within the same breed, and some characteristics may have slight variations among individual birds.

Below is a list of some popular rare and unique chicken breeds along with their egg colors, personality, size, climate tolerance, and typical use.

Rare Chicken Breeds

Chicken Breed	Characteristics
Dorking	• Egg Color: Cream to light brown • Personality: Calm, friendly • Climate: Tolerant of various climates • Size: Large • Use: Dual-purpose
Crevecoeur	• Egg Color: White • Personality: Active, flighty • Climate: Tolerant of various climates • Size: Medium • Use: Ornamental
Scots Dumpy	• Egg Color: Cream to light brown • Personality: Docile, friendly • Climate: Tolerant of various climates • Size: Small to medium • Use: Dual-purpose
Holland Chicken (Groninger Meeuwen)	• Egg Color: Cream to light brown • Personality: Docile, friendly • Climate: Tolerant of various climates • Size: Medium • Use: Ornamental, sometimes used for egg production

Brakel	• Egg Color: Cream to light brown • Personality: Active, flighty • Climate: Tolerant of various climates • Size: Medium to large • Use: Dual-purpose

Unique Breeds

Chicken Breed	Characteristics
Ayam Cemani	• Egg Color: Cream to light brown • Personality: Docile • Climate: Tolerant of various climates • Size: Small to medium • Use: Ornamental
Silkie	• Egg Color: Cream to light brown • Personality: Docile, friendly • Climate: Better suited for colder climates • Size: Small • Use: Ornamental, good brooders
Frizzle	• Egg Color: Light brown • Personality: Varies (depends on base breed) • Climate: Depends on the base breed • Size: Varies (depends on base breed) • Use: Ornamental, some breeds may be good egg layers
Serama	• Egg Color: Cream to light brown • Personality: Friendly, confident • Climate: Better suited for warmer climates • Size: Extremely small • Use: Ornamental

Naked Neck (Turken)	• Egg Color: Light brown • Personality: Varied (depends on base breed) • Climate: Tolerant of various climates • Size: Varies (depends on base breed) • Use: Dual-purpose
Blue Andalusian	• Egg Color: White • Personality: Active, can be flighty • Climate: Well-suited for various climates • Size: Medium • Use: Egg layer
Swedish Flower Hen	• Egg Color: Cream to light brown • Personality: Calm, friendly • Climate: Tolerant of various climates • Size: Medium • Use: Dual-purpose
Phoenix	• Egg Color: Cream to light brown • Personality: Active, good fliers • Climate: Better suited for warmer climates • Size: Medium • Use: Ornamental, sometimes used for egg production
Sultan	• Egg Color: Cream to light brown • Personality: Docile, friendly • Climate: Better suited for colder climates • Size: Small • Use: Ornamental
Faverolles	• Egg Color: Light brown • Personality: Docile, friendly • Climate: Tolerant of various climates • Size: Medium • Use: Dual-purpose

Showcasing Ornamental and Specialty Breeds

Within the captivating world of ornamental and specialty chicken breeds, feathers come in a myriad of hues and personalities are as diverse as their plumage. In this enchanting realm, we meet ornamental breeds like the elegant Sebright, the flamboyant Polish, and the gentle Silkie, each a living work of art. These breeds, adorned with unique features and distinct temperaments, add a touch of charm to any coop. Alongside them stand the sturdy warriors of the specialty breeds category: the Rhode Island Red, renowned for its resilience; the Leghorn, a prolific egg-layer; and the majestic Brahma, a gentle giant among chickens. As we delve into the fascinating details of each breed, you will discover the stories behind their feathers and the invaluable roles they play in the diverse tapestry of poultry farming.

Below is a list of a few popular ornamental chicken breeds along with their egg colors, personality, size, climate tolerance, and typical use.

Ornamental Breeds

Chicken Breed	Characteristics
Polish (Poland)	<ul><li>Egg Color: White</li><li>Personality: Active, sometimes flighty</li><li>Climate: Tolerant of various climates</li><li>Size: Small to medium</li><li>Use: Ornamental, occasionally used for egg production</li></ul>
Japanese Bantam	<ul><li>Egg Color: Cream to light brown</li><li>Personality: Friendly, docile</li><li>Climate: Tolerant of various climates</li><li>Size: Very small</li><li>Use: Ornamental</li></ul>
Sebright	<ul><li>Egg Color: Cream to light brown</li><li>Personality: Active, friendly</li><li>Climate: Tolerant of various climates</li><li>Size: Very small</li><li>Use: Ornamental</li></ul>
Sultans	<ul><li>Egg Color: Cream to light brown</li><li>Personality: Docile, friendly</li><li>Climate: Better suited for colder climates</li></ul>

	<ul><li>Size: Small</li><li>Use: Ornamental</li></ul>

Each breed on this list is a testament to generations of careful breeding and dedication, resulting in birds that excel in specific traits. From the robust Rhode Island Red, prized for its hardiness and egg production, to the sleek Leghorn, known for its prolific laying abilities, they have been refined for particular qualities that make them stand out.

Specialty Breeds

Chicken Breed	Characteristics
Araucana	<ul><li>Egg Color: Blue-green</li><li>Personality: Varied (depends on lineage)</li><li>Climate: Tolerant of various climates</li><li>Size: Medium</li><li>Use: Egg layer</li></ul>
Barnevelder	<ul><li>Egg Color: Dark brown</li><li>Personality: Calm, friendly</li><li>Climate: Tolerant of various climates</li><li>Size: Medium</li><li>Use: Dual-purpose</li></ul>
Jersey Giant	<ul><li>Egg Color: Light brown</li><li>Personality: Calm, friendly</li><li>Climate: Tolerant of various climates</li><li>Size: Very large</li><li>Use: Dual-purpose</li></ul>
Campine	<ul><li>Egg Color: White</li><li>Personality: Active, sometimes flighty</li><li>Climate: Tolerant of various climates</li><li>Size: Small to medium</li><li>Use: Ornamental, occasionally used for egg production</li></ul>

Houdan	<ul><li>Egg Color: White</li><li>Personality: Active, friendly</li><li>Climate: Tolerant of various climates</li><li>Size: Medium</li><li>Use: Ornamental, sometimes used for egg production</li></ul>

Selective Breeding and Genetics

In the intricate world of poultry husbandry, selective chicken breeding and genetics are essential for shaping breed characteristics. In this section, we explore breeding principles, genetic considerations, and responsible practices. Managing breeding programs for healthy flocks is key.

Principles of Selective Breeding for Desired Traits

Selective breeding is somewhat akin to customizing your chicken squad. You have the opportunity to isolate the traits you wish to propagate, such as prolific egg-laying or vibrant plumage. Here are some fundamental guidelines for executing this process (Thrifty Homesteader, 2023):

- Determine desired traits: First, ascertain your breeding objectives. Do you seek increased egg production, larger meaty fowls, or chickens known for their amiable disposition? Clearly define your goals.

- Genealogy documentation: Maintain comprehensive records of your chickens' ancestral lineage. Record their relationships, birthdates, and noteworthy attributes. This serves as your chickens' family archive.

- Selection and elimination: Select chickens that exemplify the desired traits and remove those that fall short. This practice concentrates favorable characteristics within your flock.

- Diversity incorporation: Avoid exclusively interbreeding within a single chicken clique. Introduce new genetics from distinct lineages periodically to preserve diversity.

- Patience and consistency: This endeavor is more of a marathon than a sprint. Substantial alterations require time and unwavering commitment.

- Strategic pairing: Match your chickens based on their genetic merits. Forge partnerships between chickens whose attributes complement each other.

- Continuous monitoring: Keep tabs on the progress of your breeding efforts. Are your chickens aligning with your breeding objectives? Adapt your strategy accordingly.

- Prioritize health: Healthy chickens make for better breeding stock. Safeguard their well-being and condition for optimal breeding outcomes.

- Exercise prudence: Refrain from breeding for traits that may jeopardize the comfort or health of your chickens. Always prioritize their welfare.

- Comprehensive documentation: Maintain meticulous records of your breeding activities. Record which chickens are being paired, the target traits, and the outcomes. Think of it as chronicling your chicken adventures in a diary.

These guidelines enable you to customize your chicken squad to match your ideal team, be it prolific egg layers or stylish chickens. You are in command of the process.

Genetic Considerations and Responsible Breeding Practices

Understanding the ins and outs of chicken genetics and adhering to conscientious breeding practices form the bedrock of keeping your flock healthy while striving toward specific breeding goals. These principles are the ABCs of poultry husbandry if you are dedicated to your chickens' well-being and genetic betterment.

First things first, safeguarding genetic diversity is super important. It helps dodge health problems and keeps your flock's vitality intact. Avoid mating closely related birds to steer clear of a whole bunch of issues.

Keeping pedigrees is another big deal in responsible breeding. Documenting where your chickens come from helps you make informed decisions, prevent accidental inbreeding, and trace your flock's genetic roots.

Maintaining your breeding stock's health and resilience is also a top priority. Sick or weak birds should be left out of the breeding game to prevent passing on health problems to the next generation. So, make sure you give your chickens regular health check-ups and do not shy away from culling when necessary.

Steering clear of inbreeding is a major concern. It can lead to inbreeding depression, causing lower fertility, slower growth, and just overall poor fitness. Use genetic calculators and maybe even get some expert advice to check how closely related your birds are before setting them up for a date.

Choosing the traits you want in your flock is central to your breeding goals. Whether it is egg production, meat quality, feather color, temperament, or any other factors, having clear breeding objectives guides your selection process.

Having a structured breeding program is key. You need to carefully pick the right breeding pairs and evaluate their offspring thoroughly to make sure only the genetics of your best birds are carried to the next generation.

Genetic testing can be a handy tool for ensuring strong genes. Evaluating your chickens' genetics can give you insights into specific traits or health conditions, helping you make better breeding decisions and reduce the risk of passing on hereditary complications.

Ethics should always be part of the equation and your chickens' well-being should be at the forefront. Breeding for extreme traits that harm their health or happiness should be a big no-no.

Do not forget that learning is a never-ending journey. Keep educating yourself about poultry genetics and breeding practices. Seek advice from experienced breeders, attend workshops, and stay up to date on the latest in the field to improve your knowledge.

Lastly, breeding chickens is not a sprint; it is a marathon. Making significant genetic improvements can take generations, so patience and persistence are a must.

Managing a Breeding Program for Sustainable and Healthy Flocks

Managing a breeding program for sustainable and healthy flocks is complex but rewarding. Start by selecting breeding pairs that align with your objectives and track their performance and lineage meticulously. Through regular assessment, you can ensure that only the best birds contribute to the next generation. Maintain comprehensive records to avoid inbreeding and make informed choices.

Culling underperforming or unhealthy birds may be necessary. Remember that breeding for a sustainable and healthy flock is a long-term commitment, requiring patience, diligence, and a deep understanding of your birds.

Advanced Breeding Techniques

Selective Breeding Strategies for Desired Traits

Advanced breeding techniques have completely revolutionized poultry genetics. These methods encompass a range of tools and practices, going from genomics to artificial insemination.

Genomic selection, for instance, involves careful scrutiny of a chicken's DNA to pinpoint genes connected to desired traits. This data assists breeders in identifying birds with the highest potential for transmitting these traits to their offspring, thereby improving the precision and efficiency of breeding programs.

Artificial insemination is another valuable technique for managing a flock's genetics. It allows breeders to perform specific crosses without direct mating, which is especially advantageous for preserving rare or valuable breeds.

Embryo transfer is a cutting-edge method that makes it possible to move embryos from one hen to another. This technique is useful for propagating valuable genetics or rescuing embryos from underperforming hens.

Moreover, cryopreservation is used to freeze and store genetic material, such as sperm or embryos, for future use. This acts as a genetic repository, protecting desirable traits and enabling breeders to resurrect rare or endangered breeds.

Selective breeding strategies form the core of any successful breeding program. These tactics hinge on the careful selection of which birds to mate based on their genetic traits and performance.

Line breeding, for instance, focuses on preserving the genetics of a specific lineage or family of chickens. Birds with desirable traits are repeatedly paired within the same family, thus intensifying these traits across generations.

Crossbreeding, on the other hand, involves mating chickens from different breeds to produce hybrid vigor. This can result in offspring with enhanced growth rates, disease resistance, or other coveted attributes.

To protect genetic diversity and prevent the appearance of harmful recessive traits, it is essential to avoid excessive inbreeding. Experienced breeders know how to closely monitor pedigrees and avoid pairing closely related birds. Having a solid grasp of the genetic factors behind poultry breeding is crucial for keeping those breed standards intact and making sure your birds match up with what is expected.

Understanding Genetic Considerations and Maintaining Breed Standards

Understanding the contrast between genotype (which stands for the genetic composition) and phenotype (that refers to the observable traits) provides breeders with essential insights to make informed choices when selecting breeding pairs. This knowledge holds substantial importance for poultry breeders since it forms the basis of effective breeding programs and contributes to the maintenance of breed standards.

By recognizing that particular genes might remain unexpressed in birds' appearances but can still reside in their genetic makeup, breeders can prevent unintentional and undesirable traits in their offspring. This awareness empowers them to work towards producing birds that align with established breed standards and desired characteristics, ensuring the preservation and enhancement of specific poultry breeds.

Preserving genetic diversity is critical to prevent the negative effects of inbreeding depression. Breeders frequently employ strategies such as outcrossing or introducing new bloodlines to enhance diversity.

Adherence to breed standards is also important, especially for exhibitions or shows. These standards define the ideal physical attributes for each breed, covering characteristics from plumage color to comb type. Breeders aim to produce chickens that closely match these specifications.

Sophisticated breeding methodologies, and selective breeding strategies, along with a deep understanding of genetic considerations and breed standards, empower poultry breeders to nurture flocks displaying desired traits, genetic diversity, and compliance with breed standards. These methods play a pivotal role in the continuous improvement and preservation of poultry breeds.

Conclusion

As we conclude this comprehensive guide to responsible chicken farming, let's take a moment to reflect on the key lessons and insights you have gathered throughout this journey. From evaluating your readiness to embark on this feathered adventure to selecting the right breeds, building a secure coop, and nurturing your chickens' health, this book has been your trusted companion every step of the way.

Responsible chicken farming represents an exciting lifestyle. It is a path that reconnects us to the origin of our food, cultivates a relationship with our planet, and enhances the quality of our lives along the way.

Chicken keeping constitutes a way of life dedicated to sustainability and the welfare of your flock. Tending to chickens offers a unique opportunity to reduce your carbon footprint, commit to ethical and sustainable practices, and contribute to a greener, more compassionate future.

Chickens are remarkable creatures whose everyday habits positively impact their environment, and when you provide them with the right care, they reciprocate by enhancing your sustainability.

Alongside the pleasure of high-quality eggs and companionship, there are numerous advantages to chicken keeping, including various environmental benefits.

Your journey began with the crucial process of self-assessment. You have learned the importance of aligning your commitment level, living situation, and personal preferences with the demands of chicken raising. This self-awareness sets the foundation for your successful chicken-raising journey.

Choosing the right chicken breed is a pivotal decision. Considerations such as purpose, temperament, egg production, and breed availability guide your choices. Your takeaway: The breeds you select play a pivotal role in shaping your chicken-raising experience.

Building secure and predator-proof coops has been a recurring theme, highlighting the importance of creating safe havens for your feathered friends. You have delved into the intricacies of coop design, selecting suitable materials, understanding essential coop elements, and crafting strategies to safeguard your flock from cunning predators. Your key takeaway: A well-designed and predator-proof coop ensures the safety and comfort of your chickens.

Feeding your chickens has been a central focus. Providing proper nutrition and access to clean water are vital for their health and productivity. You have explored the dietary needs of chickens, learned about feed selection, and understood the importance of maintaining cleanliness in their feeding and drinking areas. Your takeaway: A well-fed chicken is a happy and productive one.

As you may have learned, chicken parenting is not all rainbows and roses and it comes with its own set of difficulties from dealing with other pets to addressing health issues. Many chicken keepers grapple with chicken health or behavioral matters, especially during the initial years of tending to a flock.

Throughout this journey, we have emphasized the importance of a healthy flock. You have explored common health issues, learned to perform regular health checks, and uncovered strategies for preventing and addressing chicken diseases and pests. Your key takeaway: A healthy flock is the cornerstone of successful chicken farming.

Effective communication with your flock will always be encouraged. Understanding their behavior and language is the key to fostering a harmonious and happy chicken community. By building this connection, you not only ensure their well-being but also enrich your own experience as a responsible chicken farmer.

Maximizing egg production has been a goal, and you have been equipped with strategies and techniques to encourage hens to lay more eggs. You discovered the art of creating the ideal environment, providing proper nutrition, and implementing effective management practices. Your takeaway: A few simple adjustments can yield bountiful rewards in your egg basket.

For those interested in meat production, responsible and humane practices have been at the forefront. Chapter 9, "Raising Chickens for Meat," provided insights into the humane and safe practices of harvesting and processing chicken meat. You have learned how to select the right breeds, provide proper care and nutrition, and ensure the highest standards of welfare for your chickens.

Your journey has also touched upon the critical aspect of protecting your chickens from common threats. Including effective strategies, preventive measures, and practical techniques to safeguard your flock's safety and well-being.

Responsible waste management has been underscored as an integral part of chicken farming. You have explored practical solutions to handle waste, reduce environmental impact, and harness the value of chicken manure as a valuable resource. Remember, even waste can be a valuable asset in responsible chicken farming!

Finally, you have celebrated the diversity within the chicken world. You have gained insights into various breeds and their characteristics, enabling informed choices based on your unique circumstances.

Now that you have explored the world of responsible chicken farming, it is time to roll up your sleeves and get started. With tips on self-assessment, choosing the right breeds, building coops, understanding nutrition, healthcare, and more, you are all set to dive into this exciting adventure!

With these tools, build your coop, select your breeds, and raise your chickens with care and compassion. Take pride in the sustainable practices you have embraced and share your knowledge with others. Together, chicken keepers can create a future where responsible animal husbandry enriches our lives and helps our families thrive.

Thank you for choosing this guide as your companion in the world of responsible chicken farming. Your commitment to sustainability, animal welfare, and the joys of raising chickens is a testament to the positive change we can collectively bring to our world.

Here's to a future where responsible chicken farming flourishes, enriching our lives and our planet. Happy chicken raising, and may your journey be filled with clucks of contentment and the fulfillment of a sustainable and responsible way of life.

If this book has been helpful to you, we would be grateful if you could take a moment to share your thoughts on Amazon. Your review can guide others on their chicken-raising path, offering them the wisdom and motivation to thrive. Your input strengthens our community of responsible chicken farmers, contributing to a brighter and more sustainable future for everyone involved. Thank you for being part of this journey!

Keeping the Game Alive

Now you have everything you need to raise happy and healthy chickens, it's time to share your newfound knowledge and guide other readers to the same valuable insights.

Simply by leaving your honest opinion of *Raising Chickens* by Claire Hennington on Amazon, you'll be guiding fellow chicken enthusiasts to the information they're looking for and passing on your passion for raising these feathered friends.

Thank you for your help. The world of raising chickens thrives when we share our knowledge, and you're playing a crucial role in doing just that.

[Click here to leave your review on Amazon.](https://www.amazon.com/review/review-your-purchases/?asin=BOOKASIN)

The magic of raising chickens is kept alive when we pass on our knowledge, and you're helping me to do just that.

Happy Chicken Keeping!

– Your Fellow Chicken Enthusiast, Claire Hennington

PLANS FOR BUILDING
YOUR BACKYARD
COOP

One of your first steps toward keeping chickens is to make them a home. Depending on price and aesthetics, coops can range from simple enclosures to poultry palaces. These plans offer directions for building an affordable quality coop you and your flock will love.

18 SQUARE FEET
Interior of coop

3 to 5 laying hens

About 12 eggs per week

GENERAL
Galvanized wood screws
(4) galvanized door hinges
(2) door arches
(1) board for ramp
(5-8) 1x2 boards for ramp steps
(4) 4x4 treated or cedar posts - 36"
(4) concrete footings
Paint or stain

Siding Material - Use ½" plywood or ask your local hardware store for additional options

FLOOR ASSEMBLY
(4) 2x4 - 33"
(2) 2x4 - 72"
(1¾" plywood - 72" x 36)

WALLS ASSEMBLY
(6) 2x4 - 43.25" (front wall)
(1) 2x4 - 26" (front wall)
(4) 2x4 - 72" (front and back)
(6) 2x4 - 32.25 (back wall)
(2) 2x4 - 37.25 (side walls)

ROOF ASSEMBLY
(4) 2x4 - 56"
(4) 2x4 - 18"
(2) 2x4 - 26"
(1)½" plywood - 92" x 55"

Roofing Material - 92" x 65"
Options include but are not limited to corrugated roofing or composite three-tab shingles

SIDE VIEW

Ramp
Secure roof to tops of uprights
12' x 12' opening
Hinged door

FRONT VIEW

Exterior siding
Holes can be added to front and back of roof frame to increase ventilation
Hinged door
Including framing for roof

FLOOR FRAMING

4x4 post

ROOF FRAMING

Before beginning we urge you to do your own research and ask for help. The following steps are a broad outline of how to construct a coop. Please familiarize yourself with the basics of framing walls, floors and roofs before you begin.

1 Start with the 4 posts, anchor them underground with concrete footings (see floor framing detail for measuring the placement of each post.) Then frame the floor and attach the 72" x 36" sheet of ¾" plywood to the floor frame.

2 Attach 2x4's flat on the plywood floor, flush with the floor's outer edge (see front view detail.) Then attach the 4 corner uprights (two 2x4's per upright.) Next frame the remainder of the walls following the front view detail for measurements of the front door.

3 The roof may be built separately and completed then raised onto the coop and attached at each upright. Lastly, attach the siding, install the hinged doors and the ramp, and apply stain or paint to the exterior.

Good To Know:
- All dimension lumber is 2x4 unless otherwise stated.
- One nesting box should accommodate 3-5 hens. Try attaching a wooden vegetable crate or other sturdy box in a corner about 6' to 12" off the floor. Add extra nesting boxes if needed.
- To provide inexpensive perches for nocturnal roosting, attach several sturdy natural branches across a corner of the coop.

Cedar or treated 4x4 posts in underground concrete footings

Ventilation: If you need more ventilation in the coop, add holes to the roof frame (see above.) If additional ventilation is needed you can also remove one or more of the small 2x4's in the front and back of the roof frame. Securely cover any hole or gaps with ¼" metal hardware cloth.

References

AAFCO. (n.d.). *Reading labels.* https://www.aafco.org/consumers/understanding-pet-food/reading-labels/

Amit. (2020, September 14). *Poultry farm waste disposal management.* Poultry Punch. https://thepoultrypunch.com/2020/09/poultry-farm-waste-disposal-management/

Andrews, C. (n.d.). *Deep litter in the chicken coop: Pros, cons and 4 steps to its use.* Raising Happy Chickens. https://www.raising-happy-chickens.com/deep-litter-chicken-coop.html

Arcuri, L. (2019). *This is what to know when planning a chicken coop.* The Spruce. https://www.thespruce.com/plan-and-build-your-chicken-coop-3016689

Arcuri, L. (2022, February 16). *What to know when planning a chicken coop.* The Spruce. https://www.thespruce.com/plan-and-build-your-chicken-coop-3016689

Arcuri, L. (2023, January 20). *Tips for collecting and cleaning chicken eggs.* The Spruce. https://www.thespruce.com/collect-clean-and-store-chicken-eggs-3016828

Atkins, C. (2023, April 19). *The best chicken feed for healthy feathers and plumage.* The Kansas City Star. https://www.kansascity.com/reviews/best-chicken-feed/

Azeem, S. (2023, May 28). *Are chickens loud?* (A Quick Guide). ZPoultry. https://zpoultry.com/are-chickens-loud/#How_to_Reduce_Chicken_Noises

Backyard Chicken Coops. (n.d.). *Sustainable chicken keeping for a self sufficient lifestyle.* https://www.backyardchickencoops.com.au/pages/sustainable-chicken-keeping-for-a-self-sufficient-lifestyle

Barnes, A. (2019, August 9). *Things that are toxic to chickens.* The Open Sanctuary Project. https://opensanctuary.org/things-that-are-toxic-to-chickens/

BCSPCA. (2021, September 28). *3 ways to enrich the lives of backyard chickens.* https://spca.bc.ca/news/backyard-chickens-enrichment/

Bebtoli. (2017, September 5). *Managing symptoms of poultry stress in broiler chickens.* https://www.bentoli.com/broiler-chickens-stress-management/

Better Health Channel. (2012). *Food - Pesticides and other chemicals.* Victoria State Government Department of Health. https://www.betterhealth.vic.gov.au/health/healthyliving/food-pesticides-and-other-chemicals

Brahlek, A. (2022, October 3). *How to read your feed: Chicken feed labels explained.* Grubblyfarms https://grubblyfarms.com/blogs/the-flyer/how-to-read-chicken-feed-labels?

Bri. (n.d.). *The real best chicken breeds for hot climates: Not what you've been told*. The Featherbrain. https://www.thefeatherbrain.com/blog/hot-climate-chicken-breeds

British Hen Welfare Trust. (2023, April 5). *Tips to help a broody chicken*. https://www.bhwt.org.uk/blog/health-welfare/tips-to-help-abroody-chicken/

Brook, L. (2022, June 9). *Organic agriculture helps solve climate change*. National Resources Defense Council. https://www.nrdc.org/bio/lena-brook/organic-agriculture-helps-solve-climate-change

Caughey, M. (n.d.). *Cold-weather chickens for chilly climates*. HGTV. https://www.hgtv.com/outdoors/gardens/animals-and-wildlife/cold-weather-chickens-for-chilly-climates

The Chicken Chick. (2016, March 18). *Buying chickens to start a laying flock*. https://the-chicken-chick.com/buying-chickens-to-start-laying-flock/

Chicken Whisperer Magazine. (n.d.). *Benefits & challenges of raising chickens in rural areas*. https://chickenwhisperermagazine.com/the-chicken-movement/benefits-challenges-of-raising-chickens-in-rural-areas

Cocoon Chicken Coops. (2022, February 24). *Do chickens need a light in their coop*. Cocoon. https://www.chickencoopsandhouses.co.uk/blog/do-chicken-coops-need-sun-or-shade/

Country Life Natural Foods. (n.d.). *Raising backyard chickens - The pros and cons you need to consider*. https://countrylifefoods.com/blogs/country-life-feed/raising-backyard-chickens-the-pros-and-cons-you-need-to-consider

Cunningham, D., Ritz, C., & Merka, W. (2003, July 30). *Best management practices for storing and applying poultry litter*. University of Georgia Poultry Science. https://www.thepoultrysite.com/articles/best-management-practices-for-storing-and-applying-poultry-litter

Damerow, G. (2022, June 23). *12 tips for introducing new chickens to your flock*. Tractor Supply Co. https://www.tractorsupply.com/tsc/cms/life-out-here/the-coop/chick-care/how-to-introduce-new-chickens?

Dodrill, T. (2017, November 16). *The top 10 quiet chicken breeds you should raise*. New Life on a Homestead. https://www.newlifeonahomestead.com/top-10-quiet-chicken-breeds-raise/

Food Safety and Inspection Service. (2013). *Chicken from farm to table*. USDA. https://www.fsis.usda.gov/food-safety/safe-food-handling-and-preparation/poultry/chicken-farm-table

Freedom Ranger Hatchery. (n.d.). *How to prevent and treat the 5 most common chicken diseases*. https://www.freedomrangerhatchery.com/blog/how-to-prevent-and-treat-the-5-most-common-chicken-diseases/

Freeman, P. (2022, September 8). *Dos and do nots when protecting chickens from predators*. Backyard Poultry. https://backyardpoultry.iamcountryside.com/coops/protecting-chickens-from-predators/

From Scratch Farmstead. (2022, July 14). *Beginner's guide to processing chickens at home*. https://fromscratchfarmstead.com/beginners-guide-to-processing-chickens-at-home/

Garman, J. (2021, August 24). *How to compost chicken manure*. Backyard Poultry. https://backyardpoultry.iamcountryside.com/feed-health/how-to-compost-chicken-manure/

Hamelman, S. (2023, March 23). *Best chicken breeds for predators*. The Happy Chicken Coop. https://www.thehappychickencoop.com/best-chicken-breeds-for-predators/

Hamre, M. (n.d.). *Home processing of poultry*. University of Minnesota Extension. https://extension.umn.edu/small-scale-poultry/home-processing-poultry#keeping-the-carcasses-cold-831962

The Happy Chicken Coop. (2021a, March 1). *How to store your chickens' freshly laid eggs*. https://www.thehappychickencoop.com/how-to-store-your-chickens-freshly-laid-eggs/

The Happy Chicken Coop. (2021b, August 8). *4 diseases humans get from backyard chickens: Zoonotic diseases*. https://www.thehappychickencoop.com/diseases-humans-get-from-backyard-chickens/

The Happy Chicken Coop. (2022a, October 3). *Beginner's guide to raising backyard chickens*. https://www.thehappychickencoop.com/raising-chickens/

The Happy Chicken Coop. (2022b, October 31). *The complete guide to chicken parasites*. https://www.thehappychickencoop.com/guide-to-chicken-parasites/

The Happy Chicken Coop. (2022c, June 3) *Organic chicken feed: What to know before buying*. https://www.thehappychickencoop.com/organic-chicken-feed/

HappyChicken. (2022a, May 10). *The 7 best places to buy chickens*. The Happy Chicken Coop. https://www.thehappychickencoop.com/the-7-best-places-to-buy-chickens/

HappyChicken. (2022b, October 3). *Top 7 chicken feed brands*. The Happy Chicken Coop https://www.thehappychickencoop.com/top-7-chicken-feed-brands/

Hess, T. H., & Griffler, M. (2018, February 28). *Daily diet, treats and supplements for chickens*. The Open Sanctuary Project. https://opensanctuary.org/chicken-diet-and-supplements/

Holte, L. (2022, June 16). *Watering backyard chickens*. Miller Manufacturing Company Blog. https://www.miller-mfg.com/blog/watering-backyard-chickens/

Hotaling, A. (2019, August 21). *Security checklist: No chicken coop is impenetrable*. Hobby Farms. https://www.hobbyfarms.com/chicken-coop-security-impenetrable-predators-protection/

Howell, C. (2022a, October 19). *10 best chicken breeds for kids*. The Happy Chicken Coop. https://www.thehappychickencoop.com/best-chicken-breeds-for-kids/

Howell, C. (2022b, November 23). *15 best chicken breeds for confinement and small backyards.* The Happy Chicken Coop. https://www.thehappychickencoop.com/best-chicken-breeds-for-confinement/

Idea Cafe. (n.d.). *5 tips for scaling your poultry business.* https://www.businessownersideacafe.com/business_ideas/5-tips-scaling-your-poultry-business.html

IFA. (n.d.). *The best egg laying chickens: A guide to egg production.* IFA Country Stores Blog. https://grow.ifa.coop/chickens/best-egg-laying-chickens

Jacob, J. (n.d.). *Biosecurity for small poultry flocks.* Poultry Extension. https://poultry.extension.org/articles/poultry-health/biosecurity-for-small-poultry-flocks/

Jacob, J. (n.d.). *Poultry diseases affecting the respiratory system.* Poultry Extension. https://poultry.extension.org/articles/poultry-health/diseases-of-the-poultry-respiratory-system/

Keyes, G. (2020, April 1). *Chicken training using positive techniques.* Hobby Farms. https://www.hobbyfarms.com/chicken-training-positive/

Khan, S. (2022, February 21). *How to ventilate a chicken coop: right airflow and ventilation.* East Man Egg. https://eastmanegg.com/how-to-ventilate-a-chicken-coop/

Kyakuwaire, M., Olupot, G., Amoding, A., Nkedi-Kizza, P., & Ateenyi Basamba, T. (2019). How safe is chicken litter for land application as an organic fertilizer?: A review. *International Journal of Environmental Research and Public Health, 16*(19). https://doi.org/10.3390/ijerph16193521

Landsberg, G., & Denenberg, S. (2022, October). *Social behavior of chickens.* Merck Veterinary Manual. https://www.merckvetmanual.com/behavior/normal-social-behavior-and-behavioral-problems-of-domestic-animals/social-behavior-of-chickens

Lehr, A. (2022, September 23). *How big of a chicken coop do I need?* Grubblyfarms https://grubblyfarms.com/blogs/the-flyer/how-big-of-a-coop-do-i-need

Lesley, C. (2020a, May 4). *Complete guide to egg bound chickens (symptoms, treatment and more...).* Chickens And More. https://www.chickensandmore.com/egg-bound-chicken/

Lesley, C. (2020b, May 4). *The definitive guide to egg laying problems.* Chickens And More. https://www.chickensandmore.com/egg-laying-problems/

Lesley, C. (2021, January 17). *The complete guide to chickens and water.* Chickens And More. https://www.chickensandmore.com/chickens-and-water/

Linden, J. (2015, May 19). *Management guide for the backyard flock.* The Poultry Site. https://www.thepoultrysite.com/articles/management-guide-for-the-backyard-flock

Lisa. (n.d.). *5 simple steps to aid in administering medication to your chickens*. Grit. https://www.grit.com/animals/5-simple-steps-to-aid-in-administering-medication-to-your-chickens/

Lobermeier, K. (n.d.). *How to butcher and process chickens*. Under a Tin Roof. https://underatinroof.com/blog/2022/8/22/how-to-butcher-and-process-chickens

MannaPro. (n.d.). *6 most common chicken diseases*. https://www.mannapro.com/Top-6-Chicken-Diseases

McCrea, B., & Baker, B. (2022, November 22). *Common backyard chicken behaviors*. Alabama Cooperative Extension System. https://www.aces.edu/blog/topics/farming/common-backyard-chicken-behaviors/

Meghan, H. (2021, January 18). *Best chicken breeds for foraging & free ranging*. Meyer Hatchery Blog. https://blog.meyerhatchery.com/2021/01/best-chicken-breeds-for-foraging-free-ranging/

Mormino, K. S. (2013a, July 4). *11+ tips for predator-proofing chickens*. The Chicken Chick. https://the-chicken-chick.com/11-tips-for-predator-proofing-chickens/

Mormino, K. S. (2013b, October 30). *The deep litter method of waste management in chicken coops*. The Chicken Chick. https://the-chicken-chick.com/the-deep-litter-method-of-waste/

Mormino, K. S. (2016, January 18). *Chicken first aid: How to treat a sick or injured chicken*. Hobby Farms. https://www.hobbyfarms.com/chicken-first-aid/

Morning Chores. (2019, August 21). *11 best meat chickens to breed and raise in your backyard*. https://morningchores.com/meat-chickens/

Murana Chicken Farm. (n.d.). *How I saved $450 on chicken feed last year (you can too!)*. https://www.muranochickenfarm.com/2017/08/save-money-on-chicken-feed.html

Nakyinsige, K., Fatimah, A. B., Aghwan, Z. A., Zulkifli, I., Goh, Y. M., & Sazili, A. Q. (2014). Bleeding efficiency and meat oxidative stability and microbiological quality of New Zealand white rabbits subjected to halal slaughter without stunning and gas stun-killing. *Asian-Australasian Journal of Animal Sciences, 27*(3), 406–413. https://doi.org/10.5713/ajas.2013.13437

Nature's Best Organic Feeds. (2020, March 1). *Chicken layer crumbles vs. pellets: What's the difference?* https://organicfeeds.com/chicken-layer-crumbles-vs-pellets/

Norris, M. (2020, July 10). *How to butcher a chicken*. Melissa K Norris Blog. https://melissaknorris.com/podcast/how-to-butcher-chickens-part-2-of-raising-meat-chickens/

Nutrena. (2019). *The poultry digestive system*. https://www.nutrenaworld.com/blog/the-poultry-digestive-system

Oklahoma State University. (2021, July 15). *Jersey giant chickens*. https://breeds.okstate.edu/poultry/chickens/jersey-giant-chickens.html

Oklahoma State University. (2021a, July 19). *Leghorn chickens.* https://breeds.okstate.edu/poultry/chickens/leghorn-chickens.html

Oklahoma State University. (2021b, July 19). *Plymouth rock chickens.* https://breeds.okstate.edu/poultry/chickens/plymouth-rock-chickens.html

Oklahoma State University. (2021c, July 19). *Rhode Island red chickens.* https://breeds.okstate.edu/poultry/chickens/rhode-island-red-chickens.html

Oklahoma State University. (2021d, July 20). *Sussex chickens.* https://breeds.okstate.edu/poultry/chickens/sussex-chickens.html

Omlet US. (n.d.-a). *Laws about keeping chickens.* https://www.omlet.us/guide/chickens/laws_about_keeping_chickens/

Omlet US. (n.d.-b). *Loudest chicken breeds.* https://www.omlet.co.uk/guide/chickens/choosing_your_chickens/loudest_chicken_breeds/

Penn State Extension. (2022, January 24). *Exploring two ways to direct market in pennsylvania.* https://extension.psu.edu/exploring-two-ways-to-direct-market-in-pennsylvania

Pickett, R. (2013, March 20). *Worm bin and chicken poop compost catch.* The Permaculture Research Institute. https://www.permaculturenews.org/2013/03/20/worm-bin-and-chicken-poop-compost-catch/

Ploetz, K. (2013, October 28). *Dear modern farmer: How do I know if backyard slaughter is legal?* Modern Farmer. https://modernfarmer.com/2013/10/dear-modern-farmer-know-backyard-slaughter-legal/

Poindexter, J. (2016, November 14). *Identifying 14 common chicken predators (and how to protect them).* Morning Chores. https://morningchores.com/chicken-predators/

Purina Mills. (n.d.-a). *How egg production is affected by age.* https://www.purinamills.com/chicken-feed/education/detail/how-egg-production-is-affected-by-age

Purina Mills. (n.d.-b). *Mystery solved: Why did my chickens stop laying eggs.* https://www.purinamills.com/chicken-feed/education/detail/mystery-solved-why-did-my-chickens-stop-laying-eggs

Purina Mills. (n.d.-c). *Purina start & grow medicated chick food.* https://shop.purinamills.com/products/purina-start-grow-medicated-feed?

Purina Mills. (n.d.-d). *What can chickens eat chicken treats to feed and avoid.* Purina Animal Nutrition. https://www.purinamills.com/chicken-feed/education/detail/what-to-feed-chickens-chicken-treats-to-feed-and-avoid

Rachael. (2023a, April 2). *The complete guide to heritage chicken breeds.* Dine a Chook. https://www.dineachook.com.au/blog/the-complete-guide-to-heritage-chicken-breeds/

Rachael. (2023b, April 2). *The complete guide to heritage chicken breeds.* Dine a Chook. https://www.dineachook.com.au/blog/the-complete-guide-to-heritage-chicken-breeds/

Ranson, Z. (2021a, May 1). *The benefits of organic chicken feed.* Nature's Best. Nature's Best Organic Feeds. https://organicfeeds.com/benefits-of-organic-chicken-feed/

Ranson, Z. (2021b, September 15). *Getting chickens to like you.* Nature's Best Organic Feeds. https://organicfeeds.com/how-to-get-chickens-to-like-you/

RSPCA. (n.d.). *Do slow growing meat chickens have better welfare than fast growing meat chickens?* RSPCA: Knowledge Base. https://kb.rspca.org.au/knowledge-base/do-slow-growing-meat-chickens-have-better-welfare-than-fast-growing-meat-chickens/

Sato, Y., & Wakenell, P. (2022, October). *Physical examination of backyard poultry.* Merck Veterinary Manual. https://www.merckvetmanual.com/exotic-and-laboratory-animals/backyard-poultry/physical-examination-of-backyard-poultry

Smith, K. (2020a, June 26). *All the different types of chicken feed explained.* Backyard Chicken Coops. https://www.backyardchickencoops.com.au/blogs/learning-centre/all-the-different-types-of-chicken-feed-explained

Smith, K. (2020b, July 17). *Making best laid chicken coop plans? Here's what you need to consider.* Backyard Chicken Coops. https://www.backyardchickencoops.com.au/blogs/learning-centre/making-best-laid-chicken-coop-plans-heres-what-you-need-to-consider

Smith, K. (2020c, July 17). *Perch importance: Chickens need a good roost to sleep.* Backyard Chicken Coops. https://www.backyardchickencoops.com.au/blogs/learning-centre/perch-importance-chickens-need-a-good-roost-to-sleep

Smith, K. (2020d, July 17). *The chicken coop - Creating the best environment for your chickens.* Backyard Chicken Coops. https://www.backyardchickencoops.com.au/blogs/learning-centre/creating-the-best-environment-for-your-chickens

Smith, K. (2020e, July 21). *How to look after a broody hen who is incubating.* Backyard Chicken Coops. https://www.backyardchickencoops.com.au/blogs/learning-centre/how-to-look-after-a-broody-hen-who-is-incubating

Smith, K. (2020f, July 23). *The ultimate guide to chicken coop predator prevention.* Backyard Chicken Coops. https://www.backyardchickencoops.com.au/blogs/learning-centre/ultimate-guide-chicken-coop-predator-prevention

Spencer, L. (2021, April 1). *What do seasonal changes mean for chickens?* Barastoc Poultry. https://barastocpoultry.com.au/what-do-seasonal-changes-mean-for-chickens/

Successful Farming Staff. (2015, January 20). *Repurposed chicken coops*. Successful Farming. https://www.agriculture.com/family/living-the-country-life/repurposed-chicken-coops

Talbot, S. (2022, May 25). *Raise broilers right for great meat, good quality of life*. Hobby Farms. https://www.hobbyfarms.com/raise-broilers-right-for-great-meat-good-quality-of-life/

Thesing, G. (2019, March 22). *Chicken breeds, hybrids, crossbreeds… just what are they?* The Scoop from the Coop. https://www.scoopfromthecoop.com/breeds-hybrids-crossbreeds-just-what-are-they/

Thrifty Homesteader. (2023, June 29). *Breeding chickens: Tips and techniques for success*. https://thriftyhomesteader.com/breeding-chickens/

The Thrifty Homesteader Team (2023, August 7). *Cornish cross chicken: Secrets to successfully raising meat chicken*s. https://thriftyhomesteader.com/cornish-cross-chicken/

U.S. Department of Agriculture. (2015, October 1). *Questions and answers – USDA shell egg grading service*. https://www.ams.usda.gov/publications/qa-shell-eggs

University of Maryland Extension. (2022, June 16). *Identifying and preventing poultry predators in the mid-Atlantic region*. https://extension.umd.edu/resource/identifying-and-preventing-poultry-predators-mid-atlantic-region-1

University of Minnesota Extension. (n.d.). *Raising chickens for eggs*. https://extension.umn.edu/small-scale-poultry/raising-chickens-eggs

University of Minnesota Extension. (n.d.). *Raising chickens for meat*. https://extension.umn.edu/small-scale-poultry/raising-chickens-meat#indoor-space-and-heat-1580710

University of New Hampshire. (2019, August 6). *Internal parasites in chickens*. https://extension.unh.edu/resource/internal-parasites-chickens

USDA APHIS. (2022, August 26). *US poultry industry manual - Broilers: Brooding*. The Poultry Site. https://www.thepoultrysite.com/articles/fad-broilers-brooding

Wallace Center. (n.d.). *Grazing benefits*. Pasture Project at Wallace Center. https://pastureproject.org/about-us/regenerative-grazing-benefits/

Winger, J. (2016a, March 28). *50+ ways to use extra eggs*. The Prairie Homestead. https://www.theprairiehomestead.com/2016/03/ways-use-extra-eggs.html

Winger, J. (2016b, June 21). *Homemade chicken feed recipe*. The Prairie Homestead. https://www.theprairiehomestead.com/2016/06/homemade-chicken-feed-recipe.html

Winger, J. (2023, January 18). *Ultimate guide to chicken nesting boxes*. The Prairie Homestead. https://www.theprairiehomestead.com/2023/01/ultimate-guide-to-chicken-nesting-boxes.html

Wyss, L. (2017, May 5). *How to humanely kill a chicken*. Insteading. https://insteading.com/blog/chicken-slaughter/

You Should Grow. (2017, December 10). *How help a sick chicken — Q & A with a chicken vet.* https://youshouldgrow.com/sick-chicken/

Image References

Anh, X. T. (March 20, 2020). *Chicken hen wallpaper* [Image]. Pixabay. https://pixabay.com/photos/chicken-hen-wallpaper-animal-4949352/

Сергей. (February 2, 2023). *Chickens incubator brood* [Image]. Pixabay. https://pixabay.com/photos/chickens-incubator-brood-poultry-7763394/

Dais, W. (May 14, 2014). *Chicken coop farm chickens* [Image]. Pixabay. https://pixabay.com/photos/chicken-coop-farm-chickens-coop-343942/

Eickhoff, S. (November 17, 2020). *Hens poultry backyard* [Image]. Pixabay. https://pixabay.com/photos/hens-poultry-backyard-chickens-5753731/

Halliday, S. (January 2, 2020). *Brown hen on person's hand* [Image]. Unsplash. https://unsplash.com/photos/brown-hen-on-persons-hand-EyJjK9_g9Ic

Myznik, E. (May 28, 2020). *Red black and white rooster* [Image]. Unsplash. https://unsplash.com/photos/red-black-and-white-rooster-WDKEg5sDz0Y

Ovsyannykov, I. (August 20, 2017). *Animal, Birdcage* [Image]. Pixabay. https://pixabay.com/photos/animal-birdcage-capon-chicken-coop-2664504/

Pasja1000. (April 26, 2018). *Poultry birds farm* [Image]. Pixaby. https://pixabay.com/photos/poultry-birds-farm-animals-nature-3355566/

Sanders, T. (December 12, 2012). *Plans for building your backyard coop* [Image]. Thetanglednest.com http://thetanglednest.com/2012/12/infographic-how-to-build-our-coop/

Zilka, S. (September 23, 2020). *Flock of black and white chickens on a green grass field during daytime* [Image]. Unsplash. https://unsplash.com/photos/flock-of-black-and-white-chicken-on-green-grass-field-during-daytime-U46bGX6KRfU

www.ingramcontent.com/pod-product-compliance
Lightning Source LLC
Chambersburg PA
CBHW082145120626
46553CB00010B/2777